Energy Conservation Systems: Concepts, Technology and Applications

Energy Conservation Systems: Concepts, Technology and Applications

Edited by **Tim Kurian**

WILLFORD **P**RESS

New York

Published by Willford Press,
118-35 Queens Blvd., Suite 400,
Forest Hills, NY 11375, USA
www.willfordpress.com

Energy Conservation Systems: Concepts, Technology and Applications
Edited by Tim Kurian

© 2016 Willford Press

International Standard Book Number: 978-1-68285-072-5 (Hardback)

Printed in the United States of America.

Contents

Preface

This book has been an outcome of determined endeavour from a group of educationists in the field. The primary objective was to involve a broad spectrum of professionals from diverse cultural background involved in the field for developing new researches. The book not only targets students but also scholars pursuing higher research for further enhancement of the theoretical and practical applications of the subject.

Energy conservation has become a very significant discipline in the wake of diminishing reserves of conventional energy sources and renewable energy sources not being harvested to their fullest potential. The topics covered in this extensive book deal with the vital topics and emerging trends in different energy conservation systems such as advanced design and techniques for efficient energy conversion, reducing environmental degradation during energy production, optimizing use of traditional fuel resources, etc. This book strives to provide the readers with a thorough understanding of the subject.

It was an honour to edit such a profound book and also a challenging task to compile and examine all the relevant data for accuracy and originality. I wish to acknowledge the efforts of the contributors for submitting such brilliant and diverse chapters in the field and for endlessly working for the completion of the book. Last, but not the least; I thank my family for being a constant source of support in all my research endeavours.

Editor

The Influence of the Heat Source Temperature on the Multivane Expander Output Power in an Organic Rankine Cycle (ORC) System

Piotr Kolasiński

Department of Thermodynamics, Theory of Machines and Thermal Systems,
Faculty of Mechanical and Power Engineering, Wrocław University of Technology,
Wybrzeże Wyspiańskiego 27, 50-370 Wrocław, Poland; E-Mail: piotr.kolasinski@pwr.edu.pl

Academic Editor: Roberto Capata

Abstract: Organic Rankine Cycle (ORC) power systems are nowadays an option for local and domestic cogeneration of heat and electric power. Very interesting are micropower systems for heat recovery from low potential (40–90 °C) waste and renewable heat sources. Designing an ORC system dedicated to heat recovery from such a source is very difficult. Most important problems are connected with the selection of a suitable expander. Volumetric machines, such as scroll and screw expanders, are adopted as turbine alternative in small-power ORC systems. However, these machines are complicated and expensive. Vane expanders on the other hand are simple and cheap. This paper presents a theoretical and experimental analysis of the operation of a micro-ORC rotary vane expander under variable heat source temperature conditions. The main objective of this research was therefore a comprehensive analysis of relation between the vane expander output power and the heat source temperature. A series of experiments was performed using the micropower ORC test-stand. Results of these experiments are presented here, together with a mathematical description of multivane expanders. The analysis presented in this paper indicates that the output power of multivane expanders depend on the heat source temperature, and that multivane expanders are cheap alternatives to other expanders proposed for micropower ORC systems.

Keywords: ORC; rotary vane expander; heat source; temperature; power

1. Introduction: The Application of Volumetric Expanders in ORC Systems

Effective energy recovery from renewable and waste energy sources is one of the most important present-day problems [1]. Advanced energy systems based on local waste heat, renewables and fossil fuel resources can provide opportunity for an increase of consumers' energy safety and energy supply continuity [2]. They can also provide the chance for development of prosumers [3], which are the energy consumers able to produce energy for their own needs. However, implementation of local and prosumer energy systems requires access to the relevant energy conversion technologies.

One of the most promising energy conversion technologies is the ORC system [4]. The Organic Rankine cycle has the same configuration as the well-known classical steam Rankine cycle but uses low-boiling working media instead of water. ORC power systems may differ in power, purpose and technical configuration. Micro power (0.5–10 kW), small power (10–100 kW), medium power (100–500 kW) and large power (500 kW and more) systems are available. They can operate as [5] power plants, CHP's and multi-generation systems.

The most important problems connected with ORC system design are the selection of a suitable working fluid and expander. Currently there is a wide range of applicable working fluids available [6] and many appropriate selection methods [7]. Expander selection is mainly based on the system power and its purpose. In general two types of expanders can be applied in ORC systems: one are turbines, the other is volumetric machines.

Turbines are mainly applied in large and medium power ORC systems powered by heat sources with high thermal capacity (1 MW and more) and temperature (150 °C and more), such as large industrial waste sources [8] e.g., steam boilers or gas turbine exhaust gases. In the large power systems silicone oils (e.g., hexamethyldisiloxane (MM) or octamethyltrisiloxane (MDM)) are mainly adopted as working fluids [9]. Since large power ORC's are now well developed, most research works focus on the optimization of the design [10].

Volumetric expanders are applied mainly in micro and small power systems [6] such as domestic and agricultural plants. One of the most important problems in this case is the dynamic thermal characteristics of the heat source also characterized by small capacity, thermal power and temperature (up to 150 °C). Variation in the heat source properties has a negative influence on the continuity of the system operation. The low thermal parameters of the heat source also influence the working fluid selection—only low-boiling working fluids can be applied in this case. These working fluids include refrigerants and similar substances e.g., classical R123 ($C_2HCl_2F_3$) and R245fa ($C_3H_3F_5$), as well as new specially designed fluids e.g., R1234yf ($C_3F_4H_2$) and R1234ze ($C_3F_4H_2$). As described above, the specific heat source characteristics result in a small working fluid flow and difficulties in system adjustment. Therefore the design and construction of small and micro power ORC systems is very difficult and most of the existing systems are still at the prototype level or under research. In low and micro power ORCs the applicability of turbines is very limited due to the machine operational characteristics requiring large working medium flows. Microturbines are very small in dimensions and there is a necessity of a very precise parts fitting what can result in high manufacturing costs. Also, the machines require very high rotational speeds. This causes difficulties in rotor balancing and the bearings. Moreover, turbine efficiency decreases with decreasing power. This is contrary to the aim of small ORC systems which should be simple, cheap and easy to use. Despite this, a few attempts have been made to apply micro-turbines in agricultural ORC system [11].

Volumetric expanders are a good option for systems where low pressures and low working medium flows are expected. In general piston, screw, spiral, vane and the rotary lobe expanders can be applied in small and micro ORC plants. The general advantages of volumetric expanders compared to turbines are:

- Small and very small capacity (volumetric expanders are able to operate with very low working medium flows),
- Low cycle frequency (which enables the quasi-static description of the operation process),
- Low specific speed,
- High pressure drops in one stage,
- Low rotational speed (approx. 3000 rev/min),
- Ease of hermetic sealing (the simplicity of the design, relatively low operational pressures and low rotational speeds results in simple machine sealing with standard seals e.g., O-rings).

The disadvantages are:

- Internal friction, reducing the efficiency and reliability of the machine,
- Need for lubrication and frequent replacement of wearing parts,
- Large weight in relation to the power and efficiency (especially in the case of piston and screw expanders),
- Internal and external leakages reducing the efficiency.

Modified and specially adapted spiral and screw compressors are often applied as expanders in micro and small ORC systems [6]. These machines are complicated to design and manufacture and are expensive. Moreover, application of a compressor as the expander results in non-optimal working conditions and lowered efficiency. To the knowledge of the author there are no specially designed ORC screw expanders available, and there is only a single prototype of an ORC-dedicated spiral expander [12].

Vane expanders are especially interesting for small and micro ORC systems. Currently they have become a subject of different experimental and numerical scientific analyses [13,14]. Micro vane expanders have a number of advantages in comparison with other volumetric expanders. The most important are:

- Very simple design,
- High power in relation to the dimensions,
- Suitability for wet gas conditions,
- Low weight,
- Negligible clearance volume,
- Ease of gas-tight sealing,
- Very low price.

There are no ORC-dedicated vane expanders available, however, standard pneumatic air motors can be easily adapted for this task.

Encouraged by the abovementioned advantages of vane expanders, the author decided to examine their practical application in a micro-ORC system. For this purpose a test-stand, described in detail later in this article, was designed and set up. The test-stand design allows research into different operating conditions due to its adaptive heat source. A series of experiments was performed using the test-stand. Analysis concerning the test-stand operation under variable amount of working fluid conditions was described by the author in his earlier article [15]. The present study concerns research related to vane expander operation under the different heat source conditions. The following aspects are treated in the following sections: the assembly and mathematical description of a rotary vane expander and the results of experiments.

2. The Assembly and Mathematical Description of a Rotary Vane Expander

As an introduction to the discussed issues the assembly and the mathematical description of a rotary vane expander is presented in the following section. Figure 1 shows a simplified scheme of the multivane rotary expander. Assembly of the expander can be described on the basis of this figure.

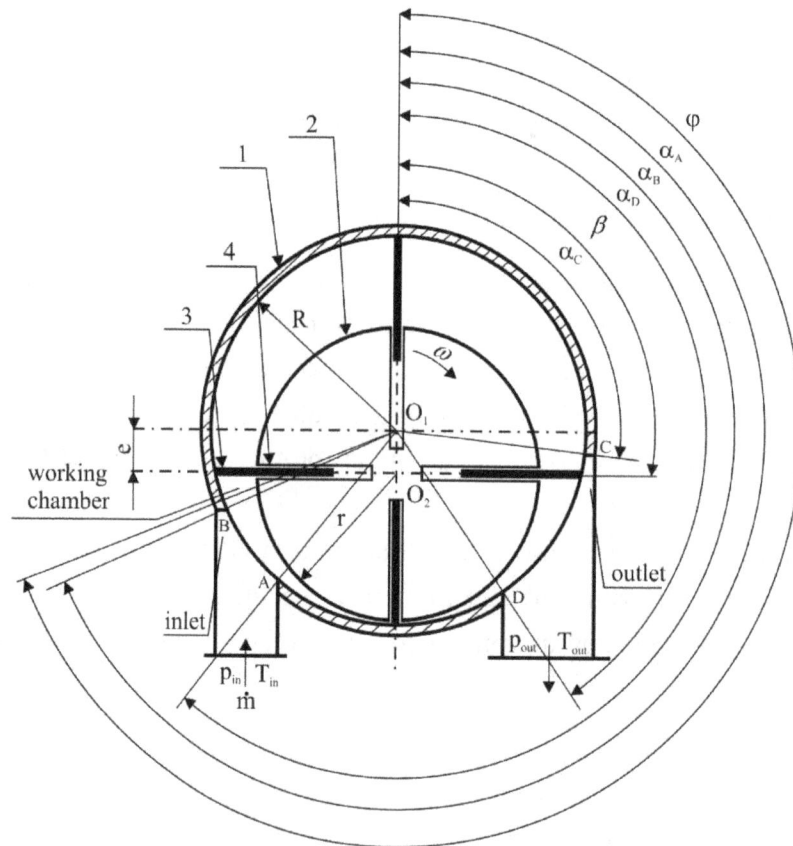

Figure 1. The scheme of the multivane rotary expander.

The main machine elements are a cylinder (1) and the rotor (2). The diameters of the rotors and cylinders may vary, depending mostly on the machine power. Small (0.02–0.05 m) diameters are characteristic of small power expanders (such as engines in small pneumatic tools, e.g., grinders or drillers), while larger diameters (0.05–0.2 m) are used in large power expanders (e.g., pneumatic engines in mixers). The rotor is mounted eccentrically in the cylinder on bearings (rolling or slide). Eccentricity can also vary, depending on the machine design. Large (0.01–0.15 m) eccentricity increases the volume of the working chambers which can improve the machine power, but can also increase the internal leakage and reduce the efficiency. Small eccentricity (lower than 0.01 m) decreases the working chambers' volume and machine power, but increases the efficiency due to the decreasing internal leakage. Vanes (3) are placed in perpendicular or inclined slots (4) milled in the rotor (see Figure 1). The vanes remain in close contact with the cylinder as a result of centrifugal force, or are pressed to the cylinder surface with help of other elements e.g., springs or rings. Inclined slots are preferable as the centrifugal force acting on the vane during rotor movement increases and the risk of jamming the vane in the slot decreases. Also, the energy dissipation by friction in the case of inclined vanes is 30%–60% smaller than in the case of perpendicular vanes [16]. The number of vanes may vary, depending on the machine design and power. More vanes can improve the continuity of the machine operation by increasing the number of coexisting working chambers, but can also decrease the efficiency by adding additional friction. The optimum is acknowledged to be between 4 and 20 vanes, but it ultimately depends on specific device design and application [16]. The cylinder is usually made of cast aluminum alloy. While the rotor is usually made of rolled steel. Vanes are made of different materials, depending on the machine type. The commonly used materials are steel, graphite, titanium and cured silicon doped with graphite. The inlet and outlet port edges (A, B, C and D in Figure 1) are referred to as the machine "steering edges". The proper arrangement of these edges has a significant influence on the expander operation. The inlet port edges (defined by α_A and α_B angles in Figure 1) and inlet port diameter should be arranged in such a way that gas can flow into the machine working chamber without significant pressure losses. The outlet port edges (defined by α_C and α_D angles in Figure 1) and outlet port diameter should be arranged in such a way that gas can be evacuated from the chamber quickly and without whirls. This issue is comprehensively described in [17].

The working principle of a multivane expander is very simple and it can be described on the basis of Figure 2. The working fluid with thermodynamic properties p_{in} and T_{in} flows into the rotating working chamber through an opening formed by the inlet port (limited by A and B) and the surfaces of the cylinder, rotor and the two vanes (V_1 and V_2) (see Figure 2a). The filling of the chamber ends when the V_2 vane passes B edge (see Figure 2b). This way, a portion of gas is enclosed within the rotating chamber. Gas pressure, exerted on the vane results in rotor movement. The vanes remain in close contact with the cylinder as a result of the centrifugal force resulting from the rotor rotational movement. During rotor movement volume of the working chamber increases and the gas expands (see Figure 2c). When the V_1 vane reaches the C edge the chamber opens and the gas starts to flow toward the outlet port (limited by C and D) (see Figure 2d). The expansion ends when the values of thermodynamic properties of the gas in the chamber equals those at the outlet port p_{out} and T_{out}. When the V_2 vane reaches the D edge, the working chamber closes again and after the minimum value of the volume is reached, the chamber is filled again with gas. The expander working cycle is completed at this point. In the multivane expander many working chambers coexists in a one moment of time and the gas is expanded concurrently in all of the chambers. This feature results in continuous machine operation.

The ideal vane expander working cycle (see Figure 4a) is formed by four ideal thermodynamic processes: isobaric filling (1–2), polytropic expansion (2–3), isobaric evacuation (3–4) and polytropic compression (4–1). The real working cycle (see Figure 4b) is different from the ideal cycle. The curved lines depicting the processes result from energy dissipation phenomena. The most important are under and overexpansion, pressure losses during filling and evacuation, friction, internal leakages, and heat transfer from the machine surface to the surroundings.

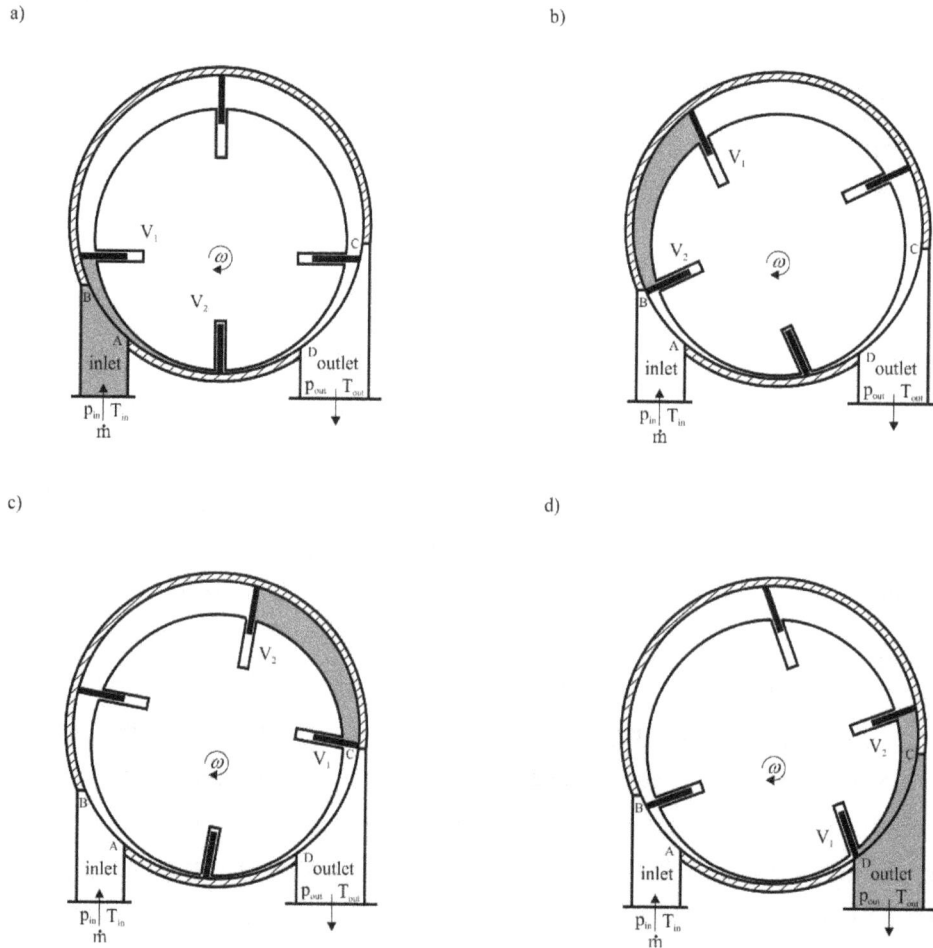

Figure 2. The process of the gas expansion in the multivane expander. (**a**) The working chamber filling; (**b**) end of the working chamber filling; (**c**) end of the expansion; (**d**) evacuation of the gas from the working chamber.

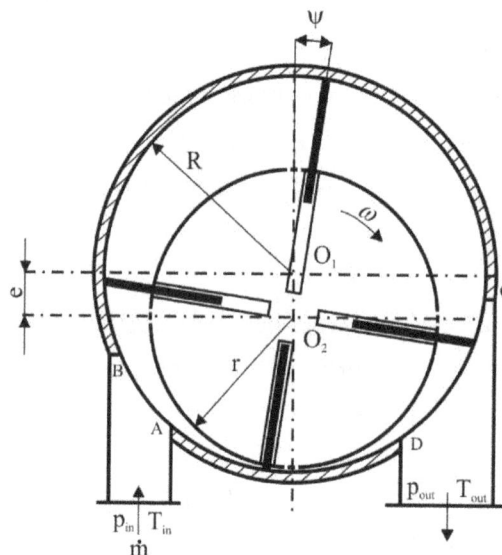

Figure 3. The scheme of the multivane rotary expander with inclined vanes.

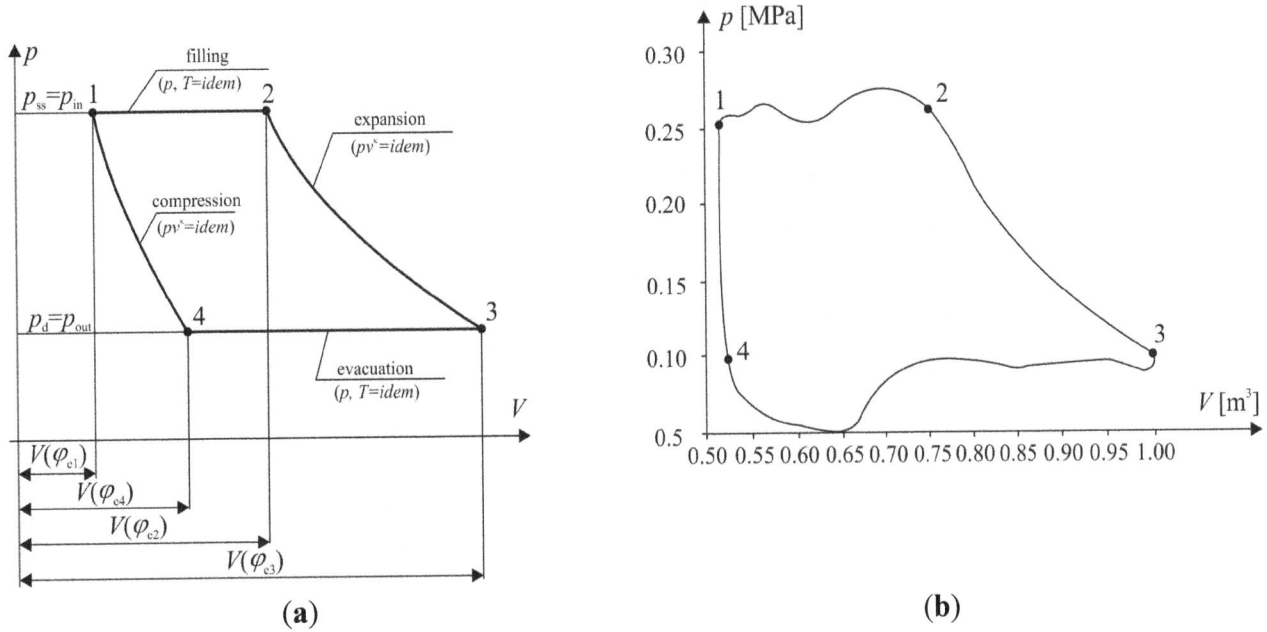

Figure 4. The comparison between the ideal and the real multivane expander thermodynamic cycles in the p-V plane [15]. (**a**) Ideal cycle; (**b**) real cycle.

In the following a brief mathematical description of the multivane expander is presented. The machine working parameters are related with the geometrical dimensions (the cylinder diameter D, the rotor diameter d, the cylinder length L, and the number of vanes z). It is generally assumed that these values are related to each other as follows [18]: $e = (0.9–0.15)R$, $L = (3–4.5)R$.

The volume of a working chamber whose position is defined by the angle φ (see Figure 1) can be described by the equation:

$$V(\varphi) = A(\varphi)L = \left(\frac{1}{2} \int_{\varphi-\frac{\beta}{2}}^{\varphi+\frac{\beta}{2}} \rho^2 d - r^2 \frac{\beta}{2} \right) L \tag{1}$$

where $A(\varphi)$ is a cross section of the working chamber, ρ is the length of the variable cylinder radius-vector, β is the angle of vane inclination to the rotor radius.

The length of the variable cylinder radius-vector whose position is defined by the angle φ can be described by the equation:

$$\rho(\varphi) = R\left(1 + \frac{e}{R}\cos\varphi - \frac{e^2}{2R^2}\sin^2\varphi \right) \tag{2}$$

where e is the eccentricity.

The combination of Equations (1) and (2) gives the following expression for the volume of a working chamber:

$$V(\varphi) = R \cdot e\left(\beta + 2\sin\frac{\beta}{2}\cos\varphi + \frac{e}{2R}\sin\beta\cos 2\varphi - \frac{e}{2R}\beta \right) L \tag{3}$$

The expansion of a gas is a polytropic process, thus the gas pressure in the chamber can be described by the equation:

$$p(\varphi) = p_{in} \left(\frac{V_{max}}{V(\varphi)} \right)^n \qquad (4)$$

where V_{max} is the maximum volume of the working chamber, n is the polytropic exponent.

Figure 5 shows the relationship between the volume of a working chamber and the angle of the rotor rotation (φ). The graph is valid under the assumption that: $R = 0.1$ m, $e = 0.09R$, $L = 3R$. Figure 6 shows the variation of pressure in a working chamber. The graph is valid under assumption that polytropic exponent $n = 1.1$ and the pressure at the expander inlet is $p_{in} = 1$ MPa.

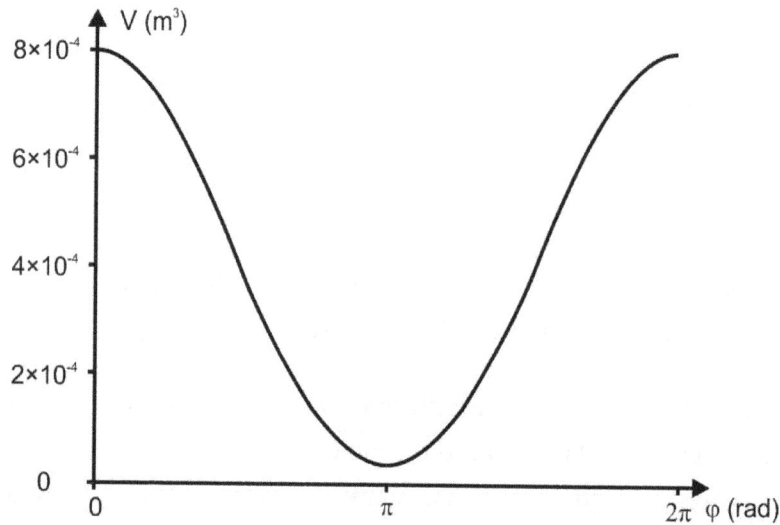

Figure 5. The relationship between the volume of a working chamber and the angle of the rotor rotation ($R = 0.1$ m, $e = 0.09R$, $L = 3R$, $\varphi = 0°–360°$).

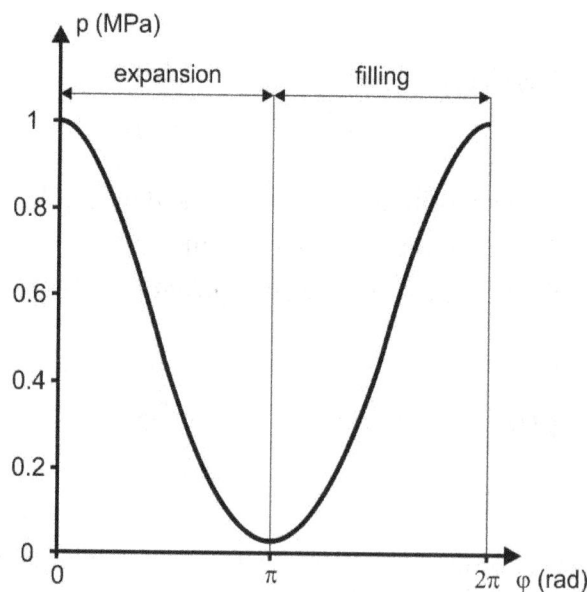

Figure 6. The variation of the gas pressure in the working chamber ($n = 1.1$, $\varphi = 0°–360°$).

The Influence of the Heat Source Temperature on the Multivane Expander Output Power...

9

The theoretical volumetric flow rate through a multivane expander can be obtained with the equation:

$$\dot{V}_t = 4\pi \cdot R \cdot e \cdot L \cdot n \tag{5}$$

where R (m) is the cylinder radius, e (m) is the eccentricity, L (m) is the cylinder length, n (rev/s) is the expander rotational speed.

The theoretical volumetric mass flow rate through a multivane expander can be obtained with the equation:

$$\dot{m}_t = \dot{V}_t \cdot \rho_{in} \tag{6}$$

where ρ_{in} (kg/m^3) is the gas density at the inlet to the expander.

The theoretical power output of a multivane expander can be obtained with the equation:

$$P_t = \dot{m}_t \left(h_{in} - h_{out} \right) \tag{7}$$

The combination of Equations (5)–(7) gives the following expression for the theoretical power output:

$$P_t = 4\pi \cdot R \cdot e \cdot L \cdot n \cdot \rho_{in} \cdot \left(h_{in} - h_{out} \right) \tag{8}$$

The volumetric flow rate through a multivane expander is dependent on the dimensionless coefficient λ representing the energy losses in the expander and it can be obtained with the equation:

$$\dot{V}_r = \lambda \cdot \dot{V}_t \tag{9}$$

λ coefficient can be obtained with the following equation:

$$\lambda = \lambda_V \cdot \lambda_T \cdot \lambda_L \tag{10}$$

where λ_V is the clearance volume coefficient; λ_T is the throttling coefficient, λ_L is the leakage coefficient. The calculation procedure for the above coefficients is very complex and has been comprehensively described in [18].

Usually, the coefficient λ is determined with the following simplified empirical equation:

$$\lambda = 1 - K \cdot \frac{p_{out}}{p_{in}} \tag{11}$$

where $K = 0.05$—for large (1 kW and more) expanders, $K = 0.1$—for small (up to 1 kW) expanders, p_{out} (Pa) and p_{in} (Pa) are the pressure of the gas at the inlet and the outlet of the expander, respectively.

The combination of Equations (8) and (9) give the following expression for the power output:

$$P_t = 4\pi \cdot R \cdot e \cdot L \cdot n \cdot \lambda \cdot \rho_{in} \cdot \left(h_{in} - h_{out} \right) \tag{12}$$

Gas can be expanded in the expander within a limited pressure range. The range can be defined by the expansion ratio:

$$\sigma = \frac{p_{in}}{p_{out}} \tag{13}$$

The multivane expander output power mainly depends on the expansion ratio [17], which is a characteristic feature of the machine and it can reach different values. Practically it is within the range of σ = 3–10 [19].

The power of a multivane rotary expander can also be obtained with the equation:

$$P = \dot{m}\left(h_{in} - h_{out}\right) \tag{14}$$

where h_{in} and h_{out} are the specific enthalpy of the working fluid at the inlet and the outlet of the expander, respectively.

The internal efficiency of a multivane rotary expander can be obtained with the equation:

$$\eta_i = \frac{h_{in} - h_{out}}{h_{in} - h_{out\,s}} \tag{15}$$

where $h_{out\,s}$ is the specific enthalpy of the working fluid at the outlet of the expander under the assumption that the expansion is isentropic.

The total efficiency of a multivane expander is dependent on many energy dissipation processes taking place in the machine during operation. The most important are [18]:

- Friction between the vane and the cylinder,
- Leakages between the working chambers,
- Heat dissipation from the casing,
- Pressure fluctuations during filling and evacuation.

According to the Reference [18] the mechanical efficiency of the multivane expanders varies from $\eta_m = 0.7$–0.75 (machines with perpendicular vanes) to $\eta_m = 0.75$–0.82 (machines with inclined vanes).

3. Description of the Experimental Test-Stand, Tested Expander and the Experiment Results

3.1. Description of the Test-Stand

An experimental test-stand was designed and realized in order to study the influence of the different operational conditions on the operation of the rotary vane expander. Figure 7 shows a simplified construction scheme of the test-stand. Figure 8 shows a general view of the test-stand.

The main test-stand components are: a gas central heating boiler (featuring maximal thermal power of 24 kW) (1), a shell-and-tube evaporator (2), a working fluid pump (3), a reservoir of working fluid (4), a plate condenser (5), and a micro multivane expander connected to a DC generator (6). The working fluid is R123. The test stand is based on manual control of the operational parameters with the help of regulation valves. The manual control helps simulate different operational conditions, e.g., it is possible to change the working medium flow direction in the evaporator from counter-flow to co-current flow. The measurements are carried out using the following methods: temperatures are measured with the use of T-type thermocouples, pressures are measured with the use of tube pressure gauges. The flow rate of R123 as well as the flow rate of cooling and heating water are measured with the use of rotameters (see Figure 7 for the measurement sensor locations: p—manometer, t—thermocouple, V—rotameter).

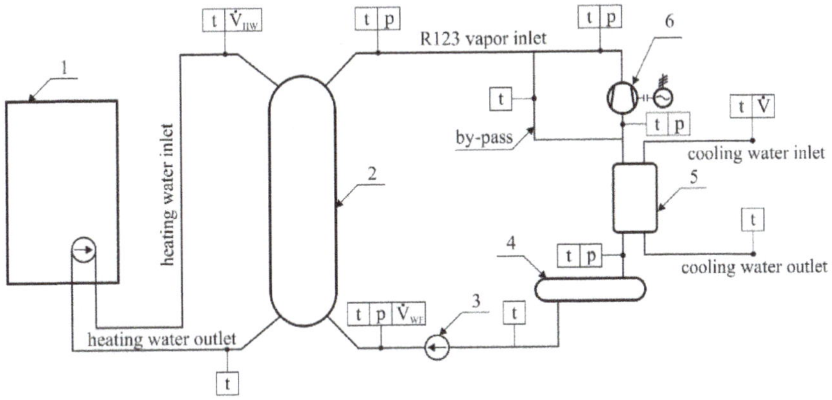

Figure 7. The simplified construction scheme of the test-stand. 1—gas central heating boiler; 2—shell-and-tube evaporator; 3—working fluid pump; 4—reservoir of working fluid; 5—plate condenser; 6—multivane expander with DC generator.

Figure 8. A general view of the test-stand [15].

The heat source for the system is hot water from the gas central heating boiler (1). The temperature of the heat source can be regulated in the range of 40–90 °C. This allows the evaluation of operational conditions of the ORC power system for variable heat sources. The working principle of the test-stand can be described on the basis of this figure (see Figure 7 for a simplified scheme of the test-stand).

The hot water from the gas central heating boiler (1) is pumped to the shell of the evaporator (2). The liquid R123 is pumped (3) from the reservoir (4) to the evaporator coil (2), where it is heated by the hot water circulating in the evaporator shell. The pressurized vapor flows through pipes to the inlet of the multivane expander (6). After the expansion the vapour flows to the plate condenser (5). The condenser is cooled by cold water in an open cycle. Then the liquid flows to the reservoir (4) and the cycle is therefore complete. A more detailed description of the test stand is presented in [20].

The expansion device is a micro four-vane air motor featuring a maximum power of 600 W. The expander was specially adapted for low-boiling working fluids, e.g., special seals and bearings were used. Moreover, a number of changes in the expander design were made in order to maximize the machine power. The expander is connected by a gas-tight clutch to a small DC generator. Figure 9 shows a view of the expander-generator unit disassembled from the test-stand. Figure 10 shows a view of expander-generator installed on the test-stand.

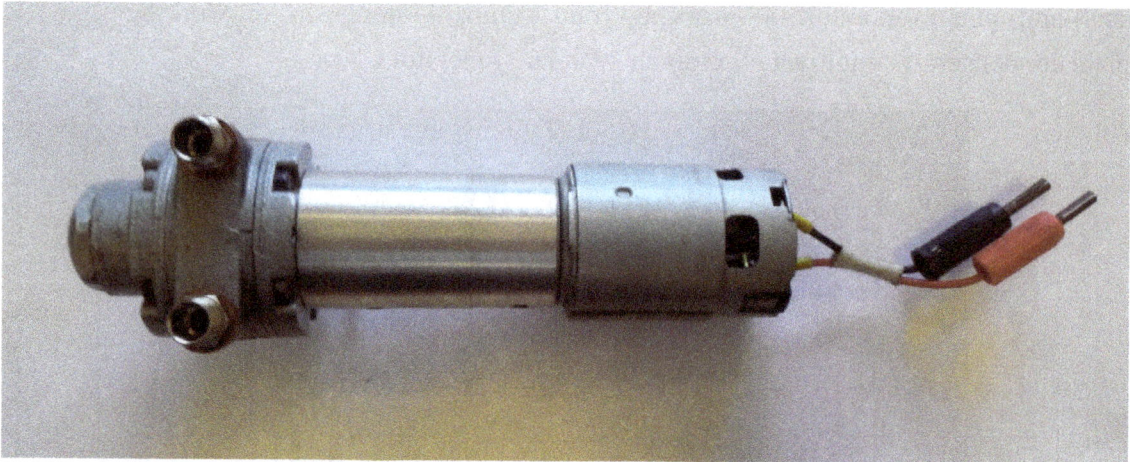

Figure 9. The view of the expander-generator unit disassembled from the test-stand.

Figure 10. The view of the expander-generator unit installed on the test-stand.

3.2. The Experimental Description and Results

The aim of the experiment is the evaluation of the multivane expander power in the variable heat source temperature case. Measurements were carried out when the system was at steady-state. Initially the working fluid pump was started and the liquid R123 was flowing through all of the test-stand components and the bypass pipeline, excluding the expander. Subsequently the central heating boiler was started. Then the water temperature setting was set on the boiler controller. The experiment was run for the water temperature range of 40–90 °C. The temperature of the water was changed by 5 °C steps, thus the measurements were made for the following values of the temperature: 40 °C, 45 °C, 50 °C, 55 °C, 60 °C, 65°C, 70 °C, 75 °C, 80 °C, 85 °C and 90 °C. The working fluid mass flow rate and the working fluid pressure were regulated by operating the regulating valve on the pump. The experiment was performed for the working fluid pressure range of 1.5–6.1 bar (6.5 bar is the maximum allowed pressure for the tested expander). When the water was heated up by the boiler to the required temperature the expander by-pass pipeline was closed by the valve. Then the vapor inlet to the expander was opened. When the system reached a steady-state the measurements of the operational parameters were made. Table 1 reports the experimental results. Table 2 shows the corresponding values of the thermodynamic properties of the working fluid at the inlet and the outlet of the expander, the expander isentropic power P_s as well as the expander power output P. The expander isentropic power was calculated using Equation (14) (with the assumption that the expansion is isentropic and $h_{out} = h_{out\,s}$).

The values of the thermodynamic properties of the working fluid were evaluated with the NIST Refprop software [21].

Table 1. Working fluid thermodynamic parameters at the inlet and the outlet of the expander.

No	t_{hs} °C	$\dot{V}_{R123}$ dm³/h	p_{in} bar	t_{in} °C	p_{out} bar	t_{out} °C
1	40	34	1.5	38.4	0.65	22.1
2	45	34	1.7	42.2	0.65	23.4
3	50	34	2.1	49.0	0.65	25.4
4	55	35	2.4	54.3	0.65	26.8
5	60	35	2.8	59.5	0.65	28.5
6	65	35	3.1	63.7	0.65	30.2
7	70	35	3.6	68.4	0.65	31.7
8	75	36	4.2	74.2	0.65	33.9
9	80	36	4.7	78.1	0.65	35.3
10	85	37	5.4	84.3	0.65	37.5
11	90	37	6.1	89.2	0.65	39.3

Figure 11 shows the test-stand thermodynamic cycle (red lines) in the T-s plane. This graph is valid for measurement no. 11 (see Table 1), where the temperature of the heat source was t_{hs} = 90 °C. Namely the thermodynamic processes presented on this figure are:

- 4–1—the evaporation of the liquid working fluid in the evaporator (element no. 2 in Figure 7),
- 1–2—the expansion of the gas in the multivane expander (element no. 6 in Figure 7),
- 2–3—the condensation of the gas in the condenser (element no. 5 in Figure 7),
- 3–4—the pumping of the liquid working fluid in the pump (element no. 3 in Figure 7).

Table 2. Thermodynamic parameters of R123 at the inlet and the outlet of the expander and values of the expander isentropic power and power output.

No	σ	h_{in} kJ/kg	s_{in} kJ/kgK	$h_{out\,s}$ kJ/kg	h_{out} kJ/kg	s_{out} kJ/kgK	x_{out}	P_s W	P W
1	2.3	405.39	1.6686	392.28	396.19	1.6821	1	184	123
2	2.6	407.76	1.6695	392.49	397.07	1.6851	1	214	145
3	3.2	411.89	1.6713	393.01	398.68	1.6905	1	264	183
4	3.7	414.83	1.6727	393.44	399.86	1.6945	1	310	215
5	4.3	417.76	1.6744	393.92	401.07	1.6985	1	346	240
6	4.8	420.09	1.6758	394.32	402.05	1.7017	1	374	262
7	5.5	422.98	1.6776	394.87	403.30	1.7059	1	408	285
8	6.5	426.43	1.6800	395.56	404.82	1.7108	1	454	318
9	7.2	428.71	1.6816	396.04	405.84	1.7141	1	480	336
10	8.3	432.10	1.6842	396.79	407.38	1.7191	1	530	371
11	9.4	434.89	1.6864	397.44	408.67	1.7233	1	562	393

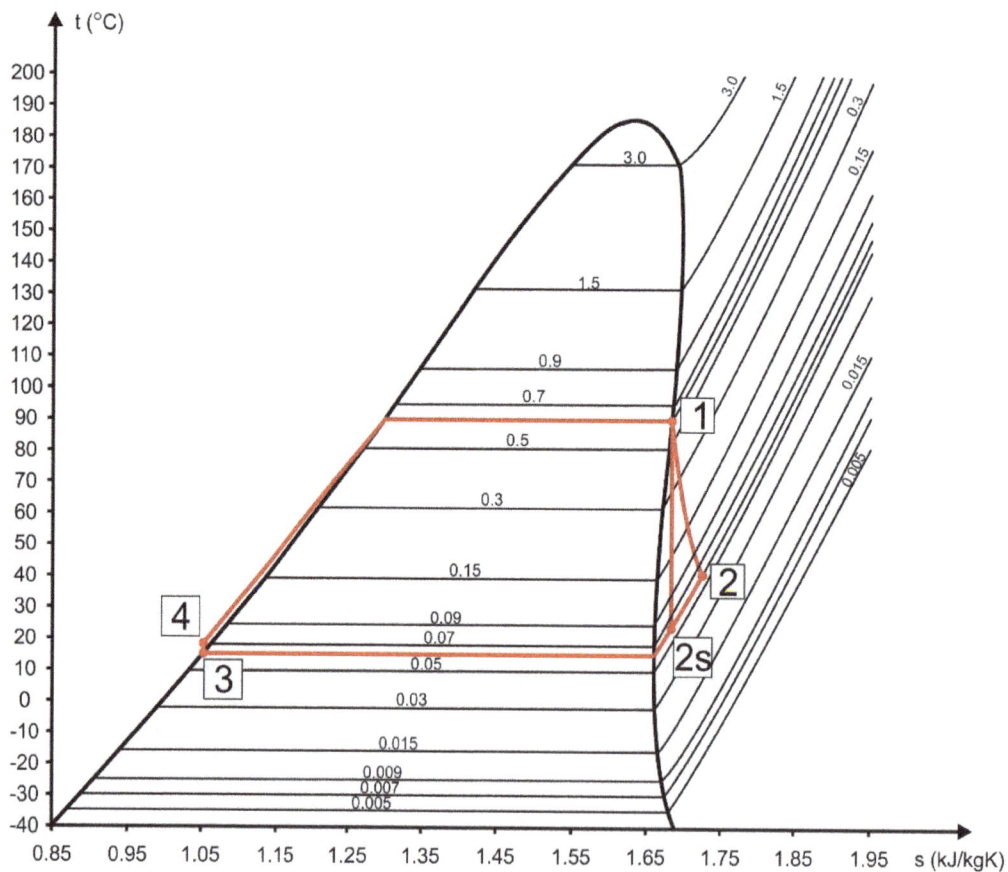

Figure 11. The test-stand thermodynamic cycle for the heat source temperature t_{hs} = 90 °C.

Figure 12 shows the variation of the expander specific work output (dashed line) during the experiment. The variation of the expander isentropic specific work (solid line) was also plotted on this graph for comparison. Figure 13 shows the variation of the expander power output (dashed line) during the experiment. The variation of the expander isentropic power (solid line) was also plotted on this graph for comparison. As it can be observed, the expander power changes depending on the heat source temperature. The minimal expander power output (123 W) was observed for the heat source temperature of 40 °C. In the next measurement points the expander power output increases with the increasing heat source temperature. The maximum expander power output (393 W) was observed for the heat source temperature of 90 °C. The power output increase is caused by the increase of the vapor pressure at the inlet to the expander. The increase of the vapor pressure is caused by the increasing heat source temperature. The effect of the working fluid flow rate changes on the expander power output is negligible as the working fluid volumetric flow rate changes during the experiment were slight (see Table 1). This is also proved by Figure 14 showing the relation between the expander power output (dashed line) and the expansion ratio. The variation of the isentropic power (solid line) was also presented on this graph for comparison. The minimal expander power output (123 W) was observed for the expansion ratio $\sigma = 2.3$, while the maximum expander power output was observed for the expansion ratio $\sigma = 9.4$. Figure 15 shows the variation of the expander internal efficiency during the experiment. As it can be observed from this graph, the expander internal efficiency varies in the range of $\eta_i = 67\%-70\%$.

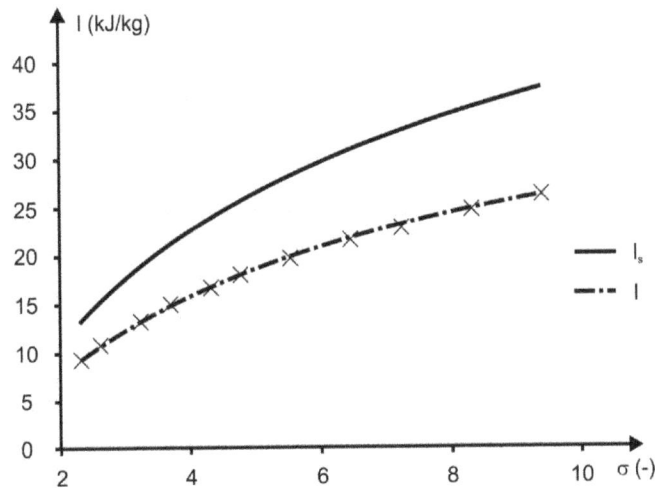

Figure 12. The variation of the expander specific work output l (dashed line, X—experimental data), the isentropic specific work l_s (solid line) and the expansion ratio.

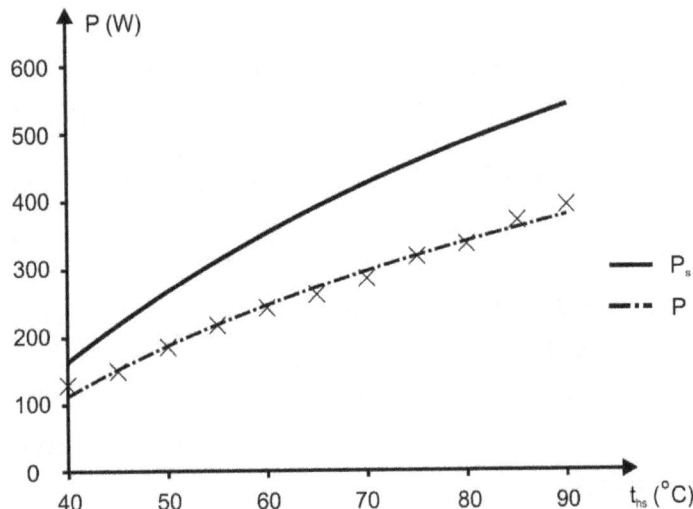

Figure 13. The variation of the expander power output P (dashed line, X—experimental data), the isentropic power P_s (solid line) and the heat source temperature.

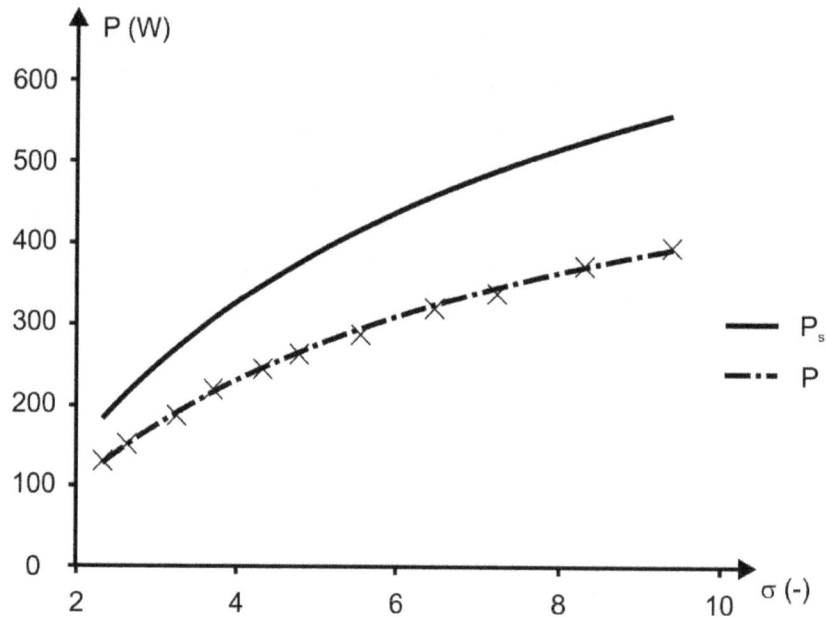

Figure 14. The relation between the expander output power P (dashed line, X—experimental data), the isentropic power P_s (solid line) and the expansion ratio.

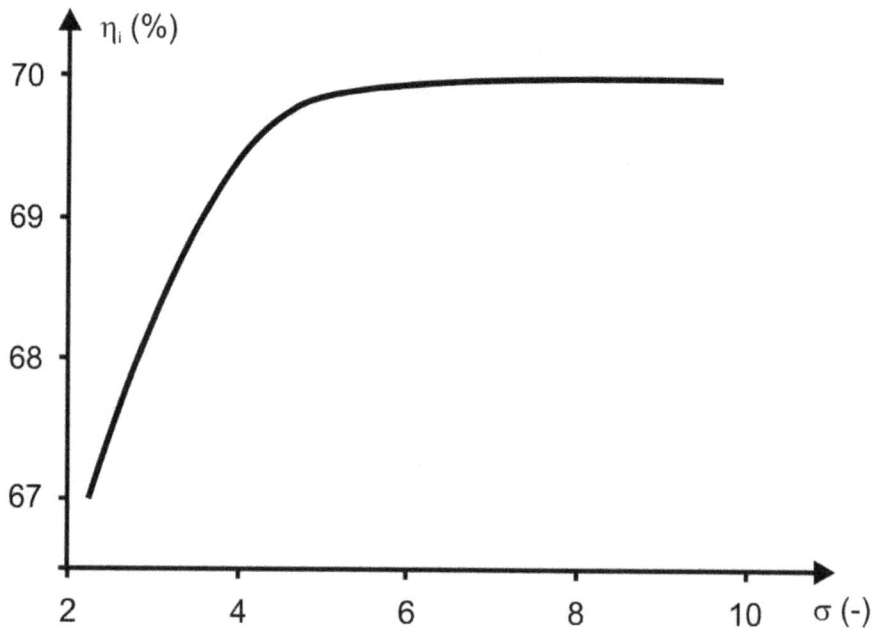

Figure 15. The variation of the expander internal efficiency and the expansion ratio.

The experimental results confirm that the output power of the rotary vane expander in an ORC system depends on the heat source temperature and expansion ratio. The working medium pressure at the inlet to the expander depends on the heat source temperature, and thus has an influence on the expander output power characteristics. During the experiments, the operation of the expander was kept under observation. It was found that the expander was working without problems, its operation was smooth and proceeded continuously.

4. Summary and Conclusions

This paper presents a theoretical and experimental analysis of operation of a the micro-ORC rotary vane expander under variable heat source temperature conditions. Such systems are suitable for dispersed cogeneration of heat and electric power in small scenarios such as households or farms.

The theoretical analysis of rotary vane expanders shows that the power output is mainly dependent on the expansion ratio (the ratio of the working fluid pressure at the inlet and outlet from the expander). The output power increases with the increasing expansion ratio. The theoretical treatment also includes a mathematical description of the expander, together with the thermodynamic descriptions. Examples of output power calculations complete the treatment.

Moreover, the experimental results related to the micro multivane expander proved that the expander power output is dependent on the heat source temperature and expansion ratio. The experiments show that commercially available multivane rotary expanders can be applied in micro ORC systems after suitable modification. The experimental results also proved that with the application of the multivane expander it is possible to recover waste heat from very low temperature (40–90 °C) heat sources. This is possible due to the multivane expander features, namely the ability to operate with low working fluid pressures and flows. This is not possible in the case of turbines and micro turbines, where high velocities and high working fluid flows are needed. Multivane expanders are also a good alternative to the other types of the volumetric expanders applied in a small power ORC systems e.g., screw, scroll and piston, as they are compact, low weight, simple in design and cheap. Moreover multivane expanders have large power output relating to their dimensions and weight when compared with earlier mentioned volumetric expanders. The experiments were performed on a very small device in a limited range of pressures. However, for larger multivane expanders, the shape of the characteristic output power, presented in Figures 12 and 13, will be similar as presented in [16]. The performance and quality of operation of the multivane expander depends on the machine design and the energy dissipation processes. The most important are internal and external leakages, internal friction, and heat transfer from the machine surface to the surroundings. Thus the design optimization of the multivane expander should include proper arrangement of the machine steering edges, and improvements to the sealing and machine insulation [17].

The application of multivane expanders can provide an opportunity for a large increase in the market share of domestic ORC systems. The simplicity of these expanders translates into easy service, operation and low system cost, what are very important in the case of domestic devices. The application of turbines in such systems is difficult and results in high manufacturing costs.

Encouraged by the advantages of these machines, the author decided to continue the research on the application of the multivane expanders in ORC systems. The author has designed a new prototype of a multivane expander-based micro ORC system combined with a heat storage system. The theoretical analysis of this system is comprehensively presented in [22]. The new prototype assembly allows one to change the working fluid. The construction of the prototype is underway. The experiment will be performed using classic working fluids i.e., R123 and R245fa, as well as modern alternatives, i.e., SES36, R1234yf and R1234ze.

Acknowledgments

The author would like to thank Roberto Capata and Terry Zhang for the invitation to publish this article.

Conflicts of Interest

The author declares no conflict of interest.

References

1. Howarth, R.B.; Andersson, B. Market barriers to energy efficiency. *Energy Econ.* **1993**, *15*, 262–272.
2. Lund, H.; Münster, E. Integrated energy systems and local energy markets. *Energy Policy* **2006**, *34*, 1152–1160.
3. Dinusha Rathnayaka, A.J.; Potdar, V.M.; Dillon, T.S.; Kuruppu, S. Formation of virtual community groups to manage prosumers in smart grids. *Int. J. Grid Util. Comput.* **2015**, *6*, 47–56.
4. Gnutek, Z.; Kolasiński, P. Organic Rankine Cycle—technology, applications and current market overview. In Proceedings of the 13th International Symposium on Heat Transfer and Renewable Sources of Energy, Międzyzdroje, Poland, 9–12 September 2010; Stachel, A.A., Mikielewicz, D., Eds.; West Pomeranian University of Technology Publishing: Szczecin, Poland, 2010; pp. 229–236.
5. Vanslambrouck, B. The Organic Rankine Cycle: Technology and applications. In Proceedings of the International Symposium: Waste Heat Recovery by ORC, Kortrijk, Belgium, Howest De Hogeschool West-Vlaanderen, 19–20 May 2009.
6. Bao, J.; Zhao, L. A review of working fluid and expander selections for organic Rankine cycle. *Renew. Sustain. Energy Rev.* **2013**, *24*, 325–342.
7. Wang, E.H.; Zhang, H.G.; Fan, B.Y.; Ouyang, M.G.; Zhao, Y.; Mu, Q.H. Study of working fluid selection of organic Rankine cycle (ORC) for engine waste heat recovery. *Energy* **2011**, *36*, 3406–3418.
8. Szargut, J. *Industrial Waste Energy—Utilization and Equipment*; WNT: Warsaw, Poland, 1993.
9. Lai, N.A.; Wendland, M.; Fischer, J. Working fluids for high-temperature organic Rankine cycles. *Energy* **2011**, *36*, 199–211.
10. Wei, D.; Lu, X.; Lu, Z.; Gu, J. Performance analysis and optimization of organic Rankine cycle (ORC) for waste heat recovery. *Energy Convers. Manag.* **2007**, *48*, 1113–1119.
11. Kozanecki, Z.; Kozanecka, D.; Klonowicz, P.; Łagodziński, J.; Gizelska, M.; Tkacz, E.; Miazga, M.; Kaczmarek, A. *Oil-Less Small Power Turbo-Machines*; Institute of Fluid-flow Machinery Publishing: Gdańsk, Poland, 2014.
12. Air Squared. Available online: http://airsquared.com/products/expanders/e25h61n50 (accessed on 17 January 2015).
13. Montenegro, G.; Della Tore, A.; Fiocco, M.; Onorati, A.; Benatzky, C.; Schlager, G. Evaluating the performance of rotary vane expander for small scale Organic Rankine Cycles using CFD tools. *Energy Proced.* **2014**, *45*, 1136–1145.

14. Montenegro, G.; Della Tore, A.; Onorati, A.; Broggi, D.; Schlager, G.; Benatzky, C. CFD simulation of a sliding vane expander operating inside a small scale ORC for low temperature waste heat recovery. *SAE Tech. Papers* **2014**, *1*, doi:10.4271/2014-01-0645.

15. Gnutek, Z.; Kolasiński, P. The application of rotary vane expanders in ORC systems—Thermodynamic description and experimental results. *J. Eng. Gas Turbines Power* **2013**, *135*, 1–10.

16. Warczak, W. *Refrigerating Compressors*; WNT: Warsaw, Poland, 1987.

17. Gnutek, Z. *Gas Volumetric Energetic Machines*; Wrocław University of Technology Publishing: Wrocław, Poland, 2004.

18. Więckiewicz, H.; Cantek, L. *Volumetric Compressors—Atlas*, 2nd ed.; Gdańsk University of Technology Publishing: Gdańsk, Poland, 1985.

19. Gnutek, Z. *Sliding-Vane Rotary Machinery. Developing Selected Issues of One-Dimensional Theory*; Wrocław University of Technology Publishing: Wrocław, Poland, 1997.

20. Kolasiński, P. Thermodynamics of Energy Conversion Systems with Variable Amount of Working Substance. Ph.D. Thesis, Institute of Heat Engineering and Fluid Mechanics, Wrocław University of Technology, Wrocław, Poland, June 2010.

21. NIST Refprop. Available online: http://www.nist.gov/srd/nist23.cfm (accessed on 17 January 2015).

22. Kolasiński, P. Use of the renewable and waste energy sources in heat storage systems combined with ORC power plants. *Prz. Elektrotech.* **2013**, *7*, 277–279.

An Electro-Thermal Analysis of a Variable-Speed Doubly-Fed Induction Generator in a Wind Turbine

Yingning Qiu [1,*], **Wenxiu Zhang** [1,†], **Mengnan Cao** [1,†], **Yanhui Feng** [1,†] **and David Infield** [2]

[1] School of Energy and Power Engineering, Nanjing University of Science and Technology, Nanjing 210094, China; E-Mails: imzhangwx@163.com (W.Z.); dl_cmn@sina.com (M.C.); yanhui.feng@njust.edu.cn (Y.F.)

[2] Department of Electronic and Electrical Engineering, Strathclyde University, Glasgow G1 1XQ, UK; E-Mail: david.infield@eee.strath.ac.uk

[†] These authors contributed equally to this work.

[*] Author to whom correspondence should be addressed; E-Mail: yingningqiu@yahoo.com or yingning.qiu@njust.edu.cn

Academic Editor: Frede Blaabjerg

Abstract: This paper focuses on the electro-thermal analysis of a doubly-fed induction generator (DFIG) in a wind turbine (WT) with gear transmission configuration. The study of the thermal mechanism plays an important role in the development of cost-effective fault diagnostic techniques, design for reliability and premature failure prevention. Starting from an analysis of the DFIG system control and its power losses mechanism, a model that synthesizes the thermal mechanism of the DFIG and a WT system principle is developed to study the thermodynamics of generator stator winding. The transient-state and steady-state temperature characteristics of stator winding under constant and step-cycle patterns of wind speed are studied to show an intrinsic thermal process within a variable-speed WT generator. Thermal behaviors of two failure modes, *i.e.*, generator ventilation system failure and generator stator winding under electric voltage unbalance, are examined in details and validated by both simulation and data analysis. The effective approach presented in this paper for generator fault diagnosis using the acquired SCADA data shows the importance of simulation models in providing guidance for post-data analysis and interpretation. WT generator winding lifetime is finally estimated based on a thermal ageing model to investigate the impacts of wind speed and failure mode.

Keywords: wind turbine; generator; lumped parameter network; thermal analysis; ageing; fault detection and diagnosis

---

1. Introduction

While wind energy development shifting from onshore to offshore, the external environment in which wind turbines (WTs) are installed has become more and more harsh [1]. Once WT components are exposed in such environments and operate in varying wind for years, it is a big challenge to achieve predictable remaining lifetime, availability and maintainability. The doubly-fed induction generator (DFIG) is one of the mainstream configurations in the wind industry, having advantages such as low cost and complexity, requiring a low converter capacity, and decoupling of active and reactive power control [2]. Temperature is one of the key parameters to define component design and materials of DFIGs [3]. A survey on wind power plants showed the major root causes for generator downtime are windings, brushes and other electrical components. The generator and electrical system cause 23.2% of WT downtime, in comparison to the gears causing 19.4% downtime [4]. Since performance and efficiency of a generator depends on the thermal condition of its windings, understanding the power losses and the thermal mechanism within a DFIG are of great importance to the industry.

In the power generation process, power losses are dissipated as heat, which directly influences the efficiency of the generator, and the temperature rise due to heat dissipation will further lower the power generator efficiency. In addition, the temperature rise within a generator will accelerate the insulation ageing and therefore affect the life of the generator. The losses related to generator include copper losses, iron losses, mechanical losses and stray losses, which are illustrated in Figure 1 [5]. Internal losses and thermal analysis within the electrical generator is necessary both for proper ventilation design and generator optimization. Copper losses and iron losses are the basic losses for a DFIG stator under normal operation conditions. The stator loss is proportional to the square of the stator current. As for most induction generators, the iron loss is less than the copper loss and it also will not be affected by the load, therefore the iron loss is neglected in many studies [6].

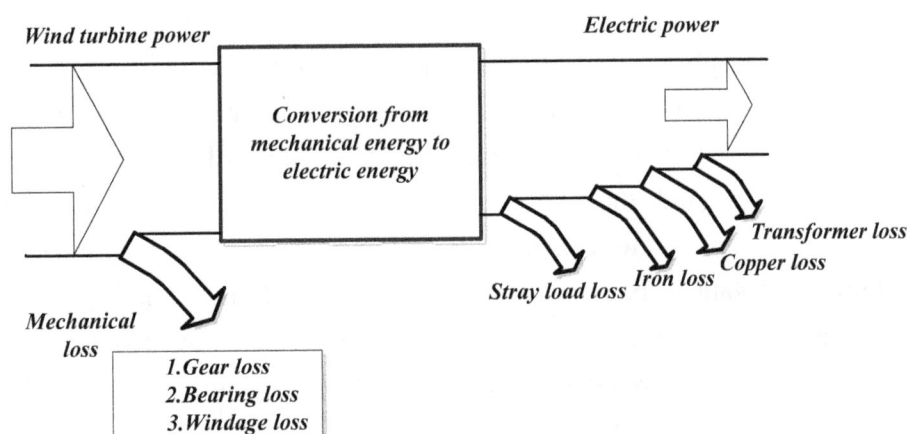

Figure 1. WT system power flow.

For a DFIG, the most important convection heat transfer occurrs between the stator and air gap, and also between the stator windings and cooling fluid [7]. Additionally, an inner fan is installed on the shaft so as to transfer the heat to the external environment. The process is illustrated in Figure 2 where the arrows represent the air flow situation.

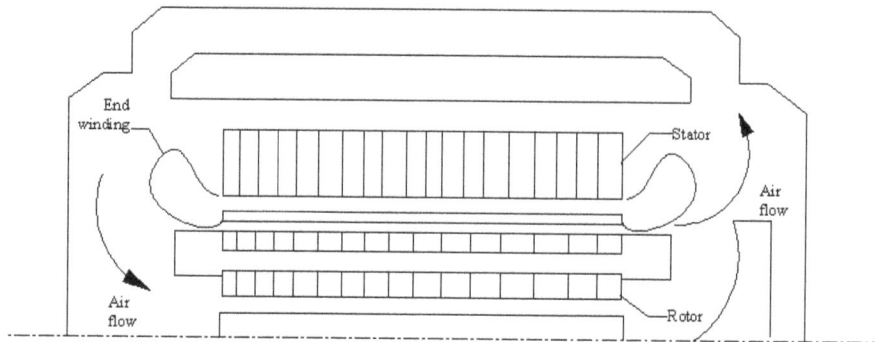

Figure 2. Sketch of the inner construction of a DFIG.

Thermal analysis of a generator is a complex problem where a number of heat exchanges are involved. Conduction, natural and forced convection and radiation are all present, and are affected by machinery electrical performance and its specific ventilation system design (natural cooling, fan cooling and water cooling, *etc.*). A coupled thermal-electromagnetic analysis approach was used to accurately predict induction motor performance. For thermal analysis, the two main approaches are analytical Lumped Parameters Network (LPN) and numerical methods. The application of these analytical techniques depends on the specific machine under study and requires considerable tailoring and modification for purpose. Commercial software packages using numerical methods based on Finite Element Analysis (FEA) or Computational Fluid Dynamics (CFD) can perform complicated multi-dimensional simulations, but they are normally time-consuming and the parameters are often difficult to determine. LPN or equivalent thermal network methods simplify the heat transfer mechanism into a thermal network with lumped parameters. The main advantages of the LPN method, such as low computational cost and capability to include the heat transfer of the cooling system, make it an appropriate method for thermal analysis of radial flux electrical generators [8].

The thermal network model has been used for transient and steady state temperature analysis of an asynchronous generator [8]. A real-time thermal model is shown for a Permanent Magnet Synchronous Motor (PMSM) which state-space format was discretized, and a model-order reduction was applied to minimize the complexity [9]. Although the LPN method has been successfully applied in induction generator thermal analysis, the aforementioned studies mostly focus on the thermophysics of generators and make the assumption of constant output torque and rotating speed. In fact, wind speed variation and control system have impacts on the thermodynamics of a WT generator. By taking into account the thermal characteristics under variable-speed operating conditions, thermal analyses should be performed to ensure the machine meets its design requirements.

Generator winding failures due to stator insulation breakdown are known as one of the most serious causes of failure. Stator insulation is affected by thermal, electrical, and mechanical stresses, as well as environmental conditions. Temporarily increasing high winding temperatures, e.g., caused by a ventilation system failure, can lead to permanent changes of the properties of the insulation system and even asphaltic run [10]. In these cases the lifetime can be reduced significantly and diagnostic tests should be performed to estimate the remaining life.

In this paper, a DFIG thermal model is firstly incorporated in a variable-speed WT model to study the thermodynamics of generator stator winding. The transient-state and steady-state temperature characteristics of stator winding under different wind speed patterns are studied. Then, applications of thermal analysis of WT DFIG on fault detection and diagnosis, and generator insulation lifetime estimation are presented. By combining both simulation and realistic data analysis, the thermal mechanisms of two failure modes, *i.e.*, generator ventilation system failure and supply voltage unbalance, are analyzed in detail to interpret the acquired data.

2. Methodology

The methodology of electro-thermal analysis of DFIG in a WT is shown in Figure 3. An electrical model is developed based on principle analysis of a wind turbine with DFIG configuration which considers the corresponding control strategy. Failure modes are then introduced into the electrical model to derive the variations of electrical parameters under fault conditions. A power loss mechanism is analyzed for the thermal model. The intrinsic thermal process of the DFIG is analyzed by an LPN model which is actually coupled to the thermal model. The variation of thermal parameters such as temperature rises of different failure modes are quantified by this electro-thermal method for fault diagnosis and thermal ageing prediction.

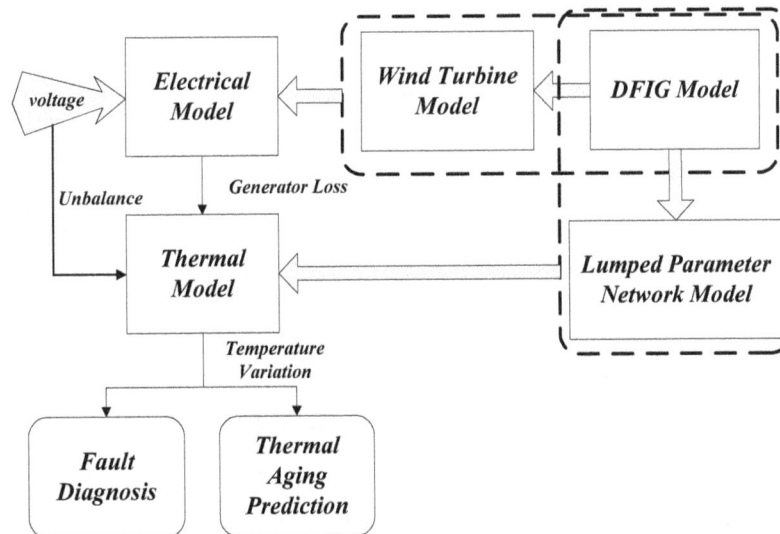

Figure 3. The methodology of an electrothermal analysis for DFIG in WT.

2.1. Wind Turbine Model

Figure 4 shows a WT with DFIG configuration. It is composed of blade, rotor, pitch system, gearbox, generator and controller. All these subsystems are controlled to realize the procedure of converting mechanical energy into electricity. The auxiliary systems in a WT include the cooling, lubrication and protection systems [11].

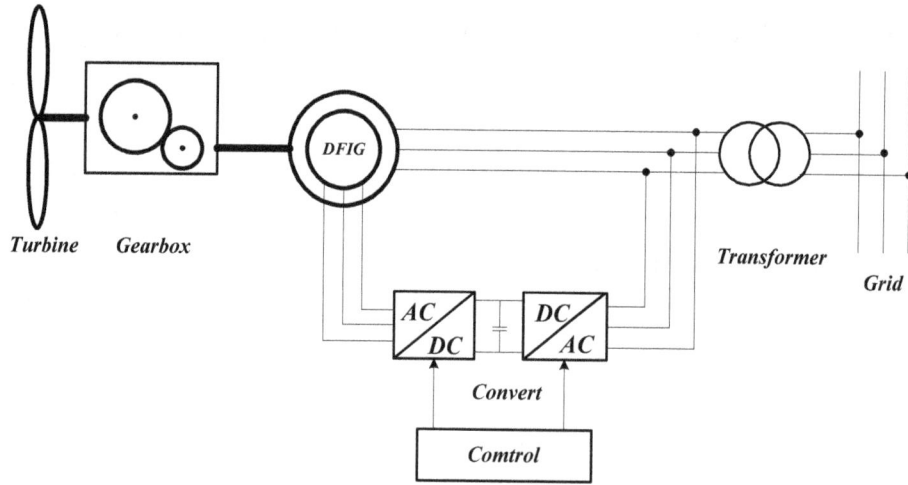

Figure 4. Configuration of a WT with DFIG.

(1) Aerodynamic principle

According to Betz theory, the power P extracted from the wind can be expressed by Equations (1)–(3) shown below:

$$P = \frac{1}{2} C_p(\lambda, \beta) \rho \pi R^2 v^3 \tag{1}$$

$$\lambda = \frac{\omega R}{v} \tag{2}$$

$$T = \frac{P}{\omega} = \frac{1}{2} C_p(\lambda, \beta) \rho \pi R^2 \frac{v^3}{\omega} \tag{3}$$

where ρ is the air density (kg/m^3), P is the derived power of WT (W), Cp is the power coefficient, λ is the tip speed ratio, β is pitch angle (rad), R is the radius of WT (m), v is the wind speed (m/s), ω is the rotor speed (rad/s), T is the rotor torque of WT (N·m).

(2) Drive train model

$$(J_r + J_g \frac{n^2}{\eta}) \frac{d\omega_r}{dt} = T_r(\omega_r, v) - \frac{n}{\eta} T_g(\omega_g, c) \tag{4}$$

Assuming a rigid drive train model, J_r is the rotational inertia of the WT rotor (kg·m^2), J_g is the rotational inertia of the generator rotors, (kg·m^2), T_r is the aerodynamic torque of the WT (N·m); T_g is the mechanical torque of the high-speed shaft (N·m) which is connected to the generator, ω_r is the rotor speed (rad/s), ω_g is the generator speed (rad/s), n is the transmission ratio of gear box, η is the efficiency of the gearbox.

2.2. Control Strategy

For a WT with DFIG, a Maximum Power Point Tracking (MPPT) control strategy is implemented in the range of wind speeds below the rated maximum by controlling the electromagnetic torque of the generator. The resultant mechanical torque that is determined by aerodynamic torque and electromagnetic torque makes the machine's rotational speed fit to a reference value in order to maintain an optimal tip-speed-ratio. In the range of wind speeds above rated, the turbine maintains constant power output by a pitch system control. Stator flux oriented vector control [12] is implemented in this study. A schematic block diagram of control system for the RSI-based current-mode control strategy is shown in Figure 5.

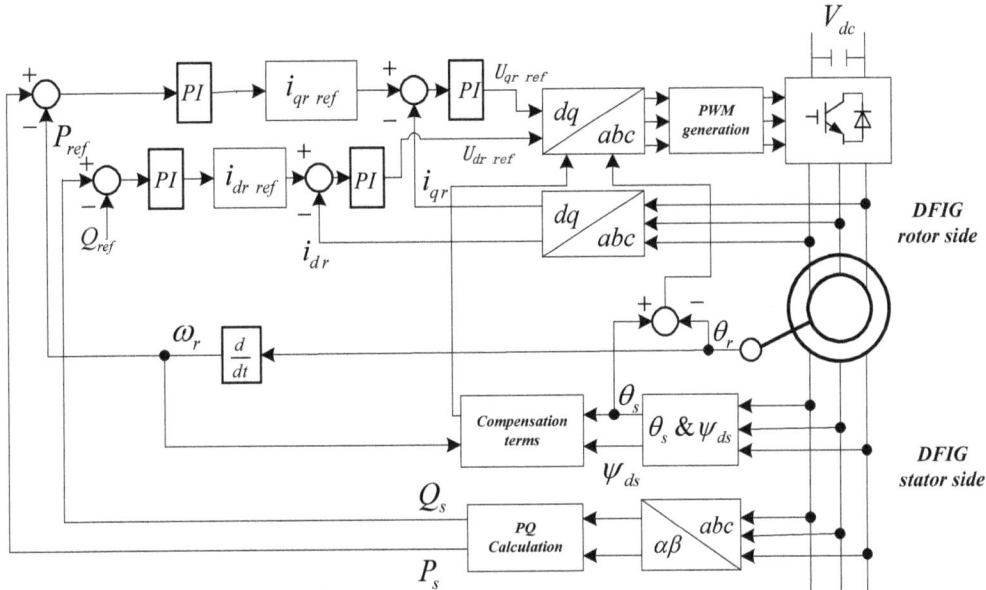

Figure 5. RSI-based stator flux vector control diagram.

The electromagnetic torque of a DFIG is controlled by the rotor current. For a DFIG rotor associated with a back-to-back converter and with the stator directly connected to the grid, the scheme controls the active and reactive power on the stator side separately. It requires the measurement of stator and rotor currents, stator voltages and rotor phase angles. The control comprises two cascaded loops: the inner loop of reactive power control and the outer loop of active power control. The optimum power is taken as the reference and it is compared with the output active power. The q-axis rotor current is controlled by a PI controller to achieve the reference of active power, which is the same for reactive power via d-axis rotor current control. Both currents are controlled by two subsequent PI controllers to determine corresponding generator rotor voltage, which will be fed back to the DFIG. Therefore, the four PI controllers are used to determine the rotor reference voltages and then the switching signals for IGBTs to produce the required voltages for converter.

2.3. Parametric Thermal Analysis

Due to the complexity of the internal structure of the generator, a very large thermal network could be constructed to obtain accurate and comprehensive temperature distributions. Due to the limitations of the thermal data acquired from the data acquisition system, a simplified thermal model of the stator windings is studied and shown in Figure 6 [7]. This model is used to predict the generator temperature variation subject to variable external wind speed and environmental disturbances.

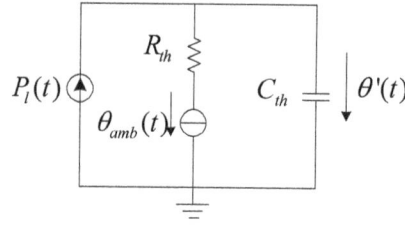

Figure 6. A generic thermal network model of the stator windings.

$P_{loss}(t)$ is the heat source determined by the stator winding power loss in the DFIG, which magnitude is determined jointly by the stator winding resistance and the current passing through:

$$P_{loss}(t) = I^2 R \tag{5}$$

I and R are the current and resistance of the generator stator winding, respectively. To consider the temperature effects to stator winding resistance, Equation (6) is used as follows:

$$R = R'(1 + \alpha \Delta T) \tag{6}$$

where R' is the initial resistance of stator winding and R is its ultimate resistance with ΔT temperature variation. α is temperature coefficient of stator winding resistance.

In the LPN model, R_{th} is the thermal resistance; C_{th} is the thermal capacitance and $\theta_{amb}(t)$ is the ambient temperature which represents the environmental variation, and $\theta'(t)$ is the generator winding temperature. Thermal resistance R_{th} is determined jointly by the thermal conduction, convection and radiation resistances. With the equivalent thermal network model established above, the thermodynamics of the stator winding are derived as follows:

$$C_{th}\frac{d\theta'(t)}{dt} + \frac{\theta'(t)}{R_{th}} = P_{loss}(t) \tag{7}$$

The transfer function is:

$$G(s) = \frac{\theta(s)}{P_{loss}(s)} = \frac{R_{th}}{1 + \tau_{th}} \tag{8}$$

where $\tau_{th} = R_{th}C_{th}$ is derived from Equation (8). The simplified thermal model is essentially a low-pass filter with a cut-off frequency of $\omega_c = 1/\tau_{th}$. The thermal resistance to model the convective heat transfer between the solid component and the fluid flow is $R_{th} = 1/(Ah)$. Failure of the ventilation system leads to the change of the heat transfer property, which is ultimately presented as a variation of the thermal parameters.

2.4. Thermal Ageing of DFIG Winding

This paper studies the reduction of stator winding insulation lifetime due to thermal stress. That is, the insulation ageing process will depend on the magnitude and duration of the operating temperature. Although thermal fatigue process is ignored in this paper, thermal ageing model is used to estimate the life of a generator with temperature fluctuation. According to the Arrhenius equation [13], the lifetime of insulation at elevated temperatures is expressed as follows:

$$\ln L = \ln B + \frac{\varphi}{kT} \tag{9}$$

where L is the life in units of time (min or hr), B is a constant, usually determined experimentally, φ is the activation energy (eV), T is the absolute temperature (K) and $k = 0.8617 \times 10^{-4}$ (eV/K) is the Boltzmann constant.

The thermal ageing model above can be modified by introducing the Half Interval Index (HIC) as insulation life is halved for every 10 °C rise in temperature. It can be written as Equation (10). This model allows estimating the insulation life for all insulation classes according to their HIC:

$$L_x = L_{100} \times 2 \exp[\frac{T_c - T_x}{HIC}] \tag{10}$$

where L_x is the lifetime percent at temperature (h), L_{100} is the lifetime percent at rated temperature (h), T_x is the hot-spot temperature for insulation class (°C), T_c is the total allowable temperature for the particular insulation class (°C), HIC is the halving interval index. For winding insulation classes A, B, F, H and H', the corresponding HIC values are 14, 11, 9.3, 8, and 10. The insulation class limit is chosen as a 20,000 h (2.3 years) period in this paper. That is, the insulation is expected to operate continuously at its maximum temperature for 2.3 years without failing.

3. Results

3.1. Thermodynamics of DFIGs in Wind Turbines

The simulation performed in this paper is based on an onshore WT equipped with a DFIG. The DFIG has two poles pairs and is air cooled. The parameters used in the simulation are listed in Table 1.

Table 1. Wind turbine parameters.

Wind Turbine	Parameter	Generator	Parameter
Rated power	850 kW	Rated voltage	690 V
Rotor radius	25 m	Frequency	50 Hz
Rated wind speed	12 m/s	Stator resistance	0.016 Ω
Cp$_{max}$	0.436	Stator inductance	6.854 mH
Lamda$_{opt}$	8	Moment of inertia	30 Kg·m^2
Air density	1.225 kg/m^3	Pole pairs	2

To obtain the generator temperature variation curve, $R_{th} = 1/25$ K/W, $C_{th} = 4{,}000$ Ws/K, ambient temperature $\theta_{amb}(t) = 20$ °C are assumed in the model [14]. The initial winding temperature is set as 0 °C and the simulation period is for 1,400 s. With the parameters above, the normalized power curve of the model WT is shown in Figure 7a. The normalized power loss and absolute stator winding temperature are obtained, and plotted against wind speed (Figure 7b,c). Power losses of the generator and the temperature of stator winding increase with wind speed.

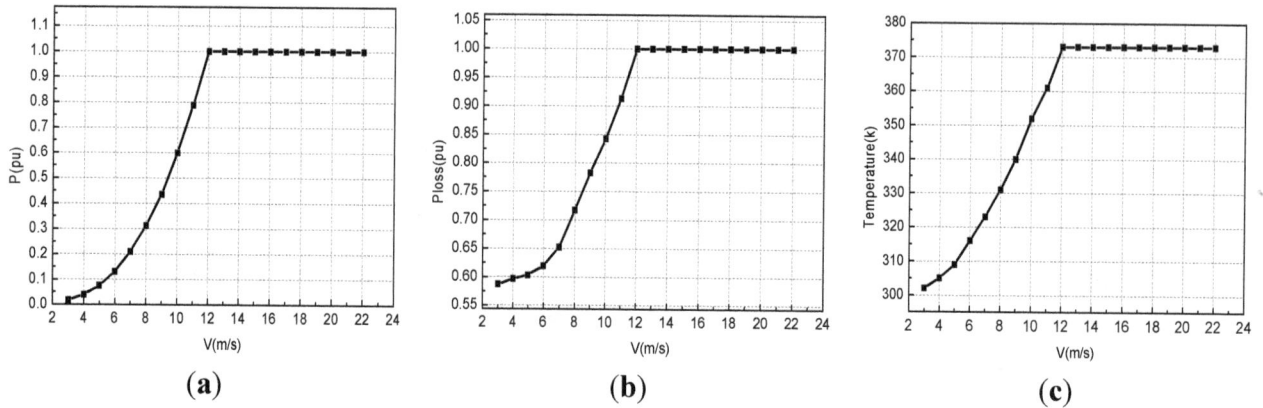

Figure 7. Power curve (**a**); Power loss against wind speed (**b**); Absolute stator winding temperature *vs.* wind speed (**c**).

The thermal behavior of the DFIG stator windings under constant wind speeds of 3, 7 and 12 m/s are obtained for a time period of 1,400 s, and shown in Figure 8. In the whole process, the generator ventilation system keeps operating continuously. From the moment that generator starts up, the stator winding experiences the first process of the machine approaching a stable operating condition and the winding reaching a constant temperature, then a second cooling down process after the machine stops at 600 s. At the higher wind speed, e.g., 12 m/s, the stable temperature reaches a higher value around 370 K due to the increase of output power. Although the thermal parameters R_{th} & C_{th} determine the time when the stator winding reaches a stable temperature, the wind speed will determine the reachable maximal temperature.

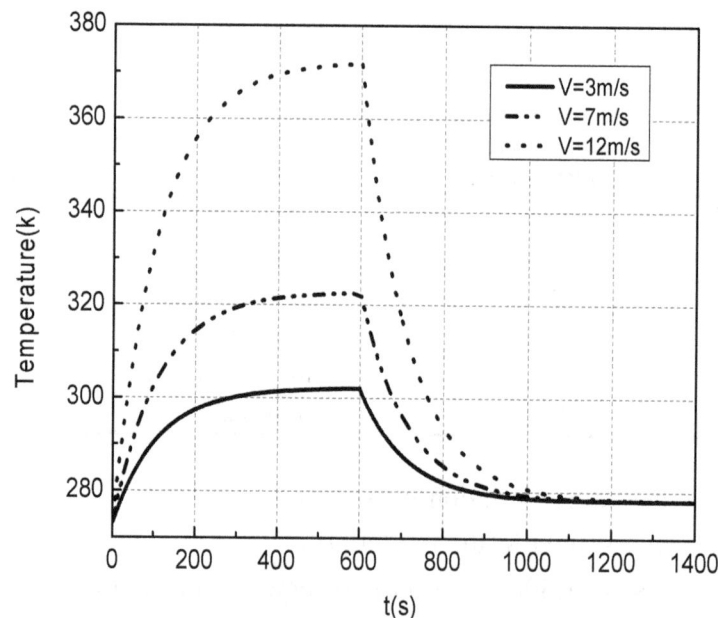

Figure 8. Thermal behaviour of the DFIG stator winding.

The mechanical, electrical and thermal responses of the DFIG under cyclic wind speed patterns are shown in Figure 9. A sudden rise of wind speed causes sharp increase of power losses, due to the transient current response in the stator winding, as shown in Figure 9d,e. However, the thermal response is not as fast as the electrical response as shown in Figure 9f. This is different from Figure 8, which shows that the stator winding temperature rise is subject to thermal inertia *i.e.*, the temperature of the stator winding takes around 600 s or 10 min to reach its maximum. This is because when the wind speed changes in a step cycle between 3 m/s (*i.e.*, start-up) and 12 m/s (*i.e.*, full power) in just 20 s, as shown in Figure 9a, before the temperature of the stator winding reaches its stabilized value, the wind speed changes cause the temperature to vary, therefore, the temperature rise in the stator winding is between 4 K to 8 K (or °C) above the ambient temperature (*i.e.*, 20 °C or 293 K) as shown in Figure 9f.

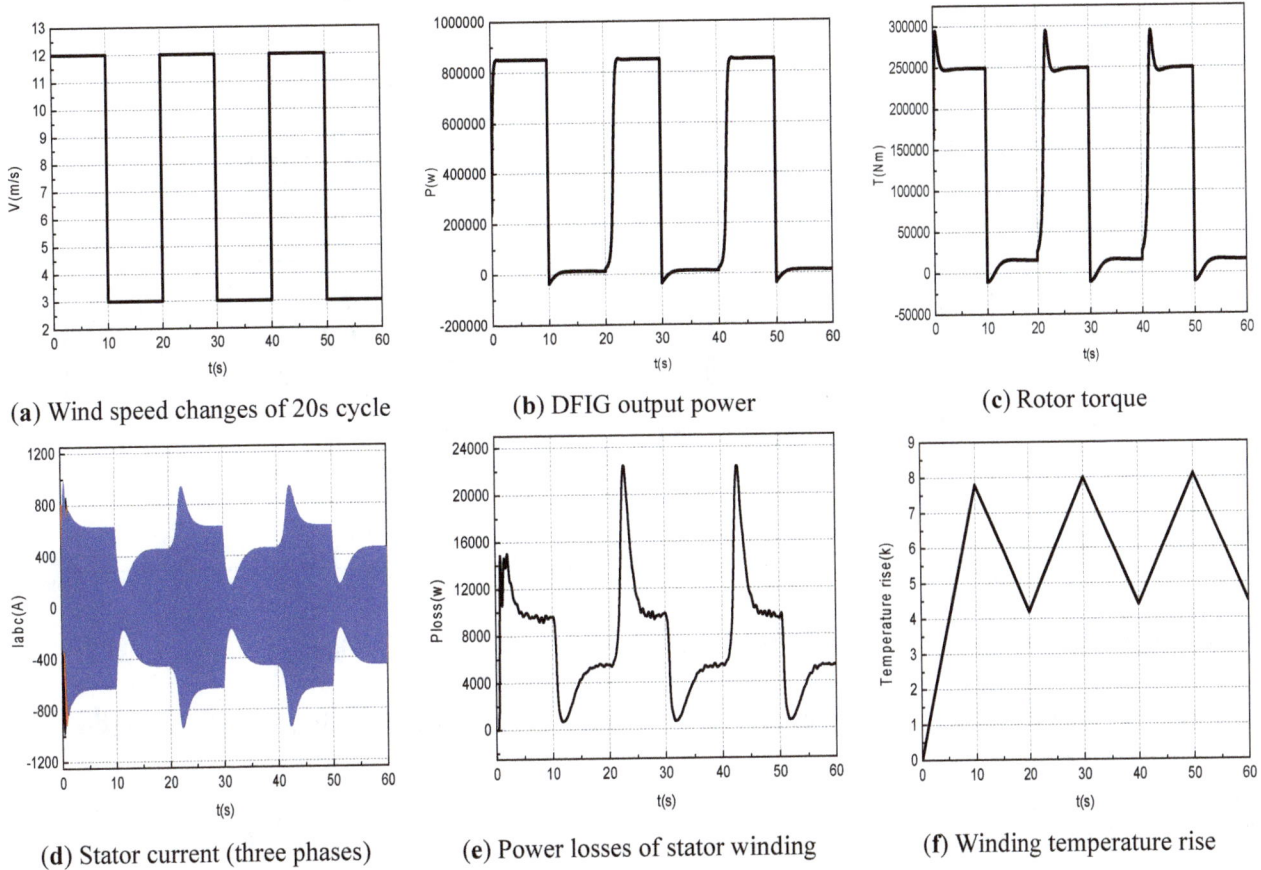

(**a**) Wind speed changes of 20s cycle   (**b**) DFIG output power   (**c**) Rotor torque

(**d**) Stator current (three phases)   (**e**) Power losses of stator winding   (**f**) Winding temperature rise

Figure 9. Mechanical, electrical and thermodynamics of the WT and DFIG stator winding under step-cycle wind speed conditions.

When the wind speed changes in a step cycles between 3 m/s (*i.e.*, start-up) and 12 m/s (*i.e.*, full power) in 30 s cycles (Figure 10) or in 40 s cycles (Figure 11), their power losses are the same but the temperature rises in the stator winding are quite different. In the 30 s case, Figure 10 shows the temperature rise is between 5 K to 10 K (or °C); In the 40 s case, Figure 11 shows the temperature rise is between 10 K to 18 K (or °C). For the same thermal parameters R_{th} & C_{th}, a DFIG machine operating with a longer high wind speed duration will experience a higher temperature rise. If the scale of the wind speed change is less than the above cases, e.g., between 3 m/s and 7 m/s, the magnitude of the temperature rise will be less. Therefore, the magnitude of temperature rise depends not only on the thermal parameters of the DFIG stator winding, but also on the wind speed profile, including the scale and cycle of the wind speed change.

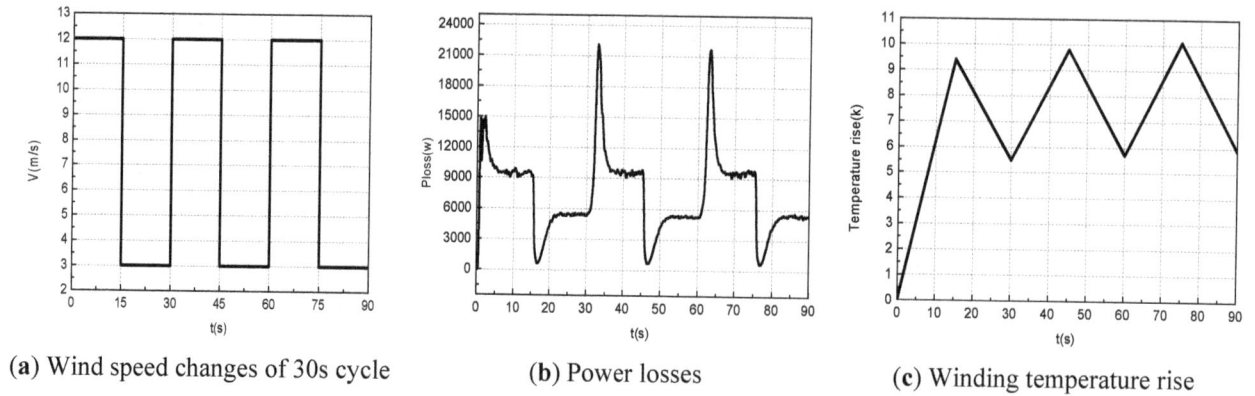

(a) Wind speed changes of 30s cycle   (b) Power losses   (c) Winding temperature rise

Figure 10. Power losses and temperature rise of stator windings for wind speed change with a 30 s cycle.

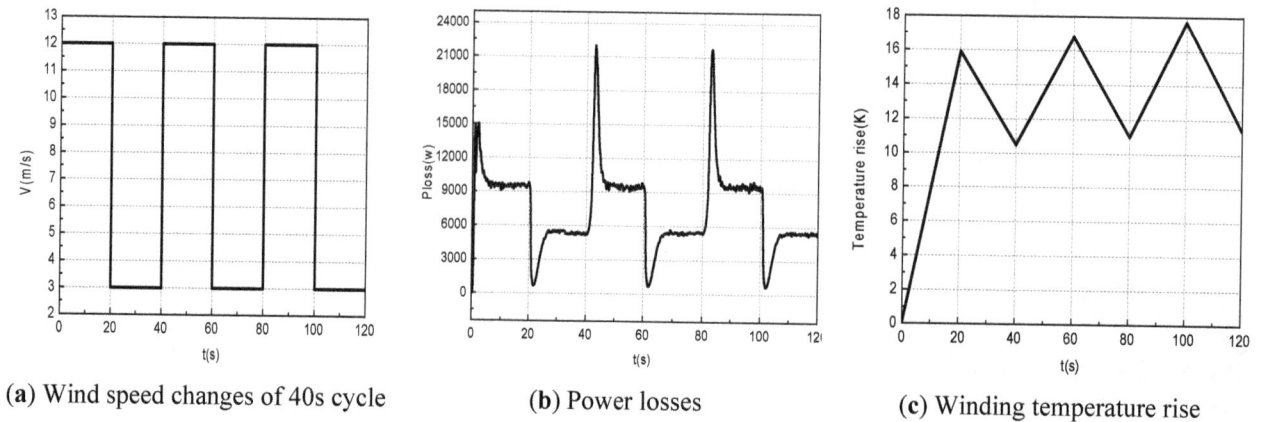

(a) Wind speed changes of 40s cycle   (b) Power losses   (c) Winding temperature rise

Figure 11. Power losses and temperature rise of stator windings a for wind speed change with a 40 s cycle.

3.2. Generator Case Studies: Voltage Unbalance vs. Ventilation Fault

Voltage unbalance, also known as the phase voltage unbalance rate (PVUR), is defined as below according to the IEEE definition:

$$PVUR = \frac{Max.\,voltage\;\;deviation\;\;from\;\;Avg.\,phase\;\;voltage\;\;magnitude}{Avg.\,phase\;\;voltage\;\;magnitude} \times 100 \qquad (9)$$

Even with the same voltage unbalance factor [15], there could be a single phase, two-phase or three-phase voltage unbalance. Here a three-phase voltage unbalance with 10% PVUR and a two-phase unbalance with 5% PVUR (b & c phases) are injected in the DFIG model to examine the corresponding thermal behaviors. The results are shown in Figure 12.

Figure 12a shows a case of 10% PVUR variation for the whole three-phase voltage on a time scale. The generator operates at the rated voltage for the first 200 s, then a 10% overvoltage for the next 200 s, and finally 10% under-voltage for the remaining 200 s. Figure 12b shows the corresponding stator winding temperature changes. Figure 12c shows the stator currents under two-phase (b & c) voltage unbalance with 5% PVUR injected at 300 s. Figure 12d shows a, b & c stator winding temperature changes. During voltage unbalance, the three-phase currents of the DFIG are non-uniform. It is quite clear that the temperature of the windings changes subject to different voltage unbalance situations.

There are several causes leading to voltage unbalance such as incomplete transposition of transmission lines, unbalanced loads, open delta transformer connections, blown fuses on three-phase capacitor hanks. Apart from the grid code violation, the voltage unbalance will also have negative impacts on electric machines, such as overheating, current unbalance, de-rating and inefficiency, leading to winding insulation degradation [16].

For a wind turbine with DFIG configuration, MPPT control is realized under the rated wind speed range to tune the rotor speed through electromagnetic torque control of the induction generator. As the induction generator torque is subject to the rotor current and stator voltage, therefore any unbalance of the stator voltage will cause an unstable rotor current in order to maintain an appropriate torque value rather than to maintain a constant power output. Therefore, an increase of stator voltage will cause a slight increase of stator current, and thus an increase of winding temperature.

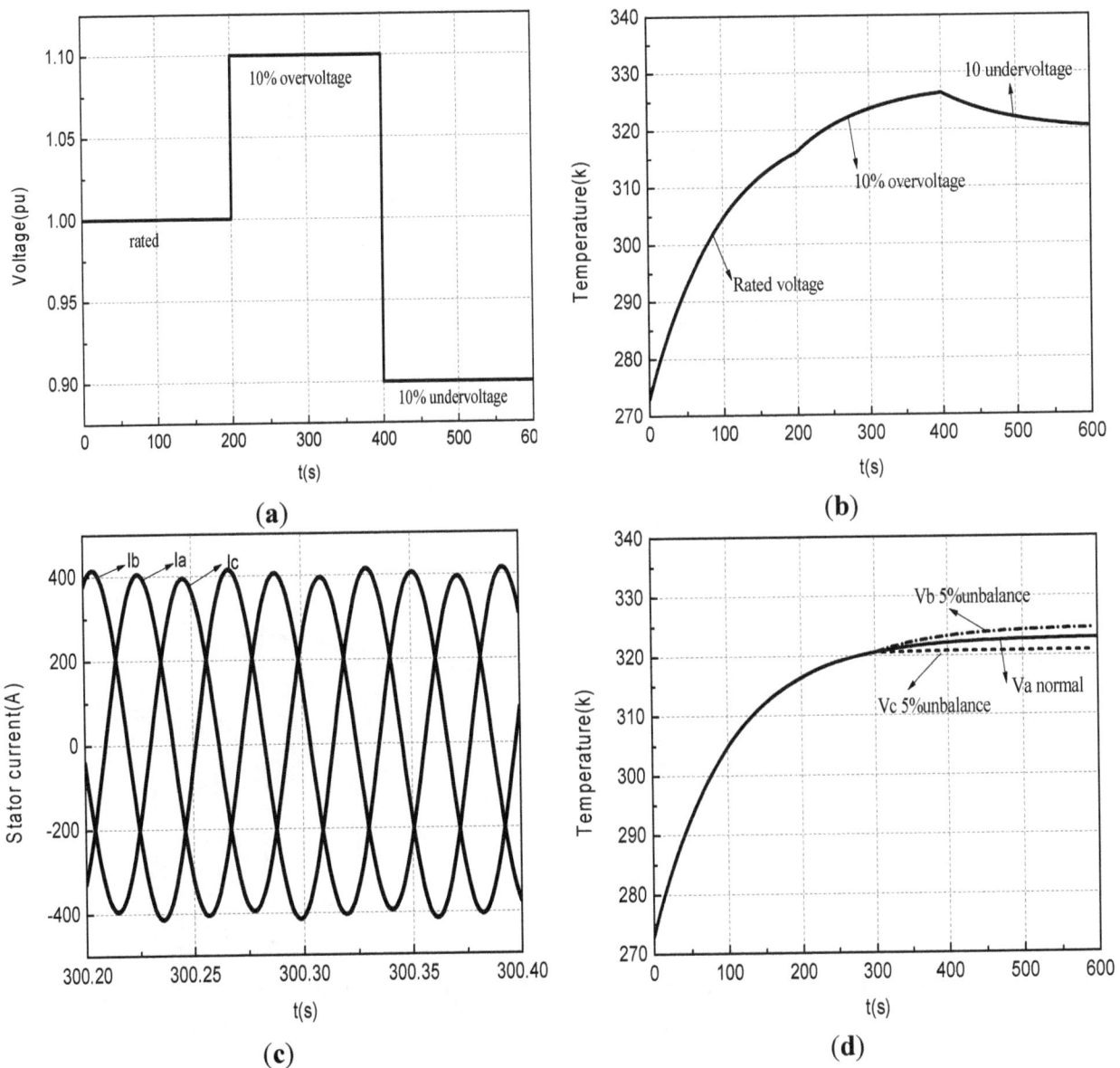

Figure 12. DFIG stator winding temperature variation due to different PVUR situations. (**a**) Three-phase voltage unbalance; (**b**) Temperature for three-phase unbalance; (**c**) b & c phase voltage unbalance; (**d**) Temperature for two-phase unbalance.

3.3. Comparisons between Simulation Results and Real WT Data

Performance data such as temperature and power output are continuously monitored by the WT SCADA system. Here real data are collected in a 10-minute interval from an operating 1–2 MW WT, and further averaged in block according to the normalized power output. The temperatures of the generator stator windings are measured by PT100 thermal resistance elements which are embedded in the windings. To reduce the data volume, the SCADA system recorded the average temperature measured from the PT100s installed in different locations of the generator stator winding. Figure 13 shows the real data collected under two conditions by plotting the absolute winding temperature rise (K) vs. normalized power output (pu) for a DFIG operating under healthy conditions (circles), and after generator ventilation failure (squares). The simulated data are further plotted under the same conditions: DFIG operating under healthy conditions (solid line), and after generator ventilation failure (dashed line). Besides the generator rating, the parameters in the simulation model such as stator winding resistance, thermal resistance and capacitance in the LPN model are adjusted accordingly to match the real data. The stator winding resistance is set to $0.016\,\Omega$ and its temperature coefficient is chosen as $0.0039\,\Omega/K$. It is clear in Figure 13 that good agreement between the simulation result and the real data is reached. From the simulation, the R_{th} is 0.0101 K/W and C_{th} is $19,200$ Ws/K before the generator ventilation failure, however, R_{th} increases to 0.0174 K/W and C_{th} remains the same after the ventilation failure. When the ventilation system is faulty, an increase of the gradient and the same intercept are found in the plot of stator winding temperature rise against power output.

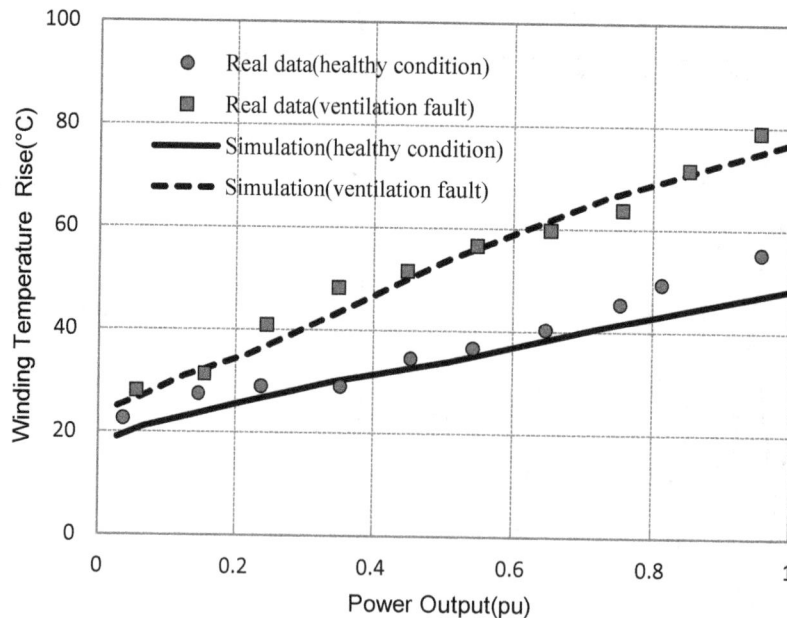

Figure 13. Winding temperature rise against normalized power output when DFIG under healthy condition and with ventilation system fault.

It is important to understand the thermal process within a DFIG generator with a faulty ventilation system. Since the generator is air-cooled, the air flow inside the generator will change from forced convection to natural convection as long as the ventilation system is faulty. In this situation, it is more difficult to dissipate the heat, which leads to an increase of the convection resistance coefficient. Due to the increase of the internal temperature of the DFIG, the stator winding resistance will increase with the temperature coefficient of 0.0039 Ω/K of copper. This further leads to an increase of power loss. The increase of power loss and the variation of thermal resistance will lead to an increase of the winding temperature rise. The fitting shows that a good agreement is obtained by increasing the thermal resistance from 0.0101 K/W to 0.0174 K/W for the DFIG switching from healthy situation to faulty ventilation circumstances. The temperature rise of the DFIG is determined jointly by the stator winding resistance and thermal resistance while the thermal capacity C_{th} remains the same. The increase of thermal resistance of R_{th} will also result in a shorter time for the generator temperature to become stable, which can be observed from the detailed analysis that shows increased scattering of the temperature data

Figure 14 shows the real data of absolute winding temperature rise (K) *vs.* normalized power output (pu) collected under another two conditions: a DFIG operating under healthy conditions (circles), and a DFIG operating under all three-phase voltage unbalance with 5% overvoltage (triangles). Similarly, the simulated data are further plotted under the same machine conditions, including a DFIG operating under healthy conditions (solid line) and a DFIG operating with a three-phase voltage unbalance with 5% overvoltage (dotted line). After the supply voltage is unbalanced with 5% overvoltage, the line intercept becomes greater, the gradient of stator winding temperature rise against power output remains the same as under healthy conditions.

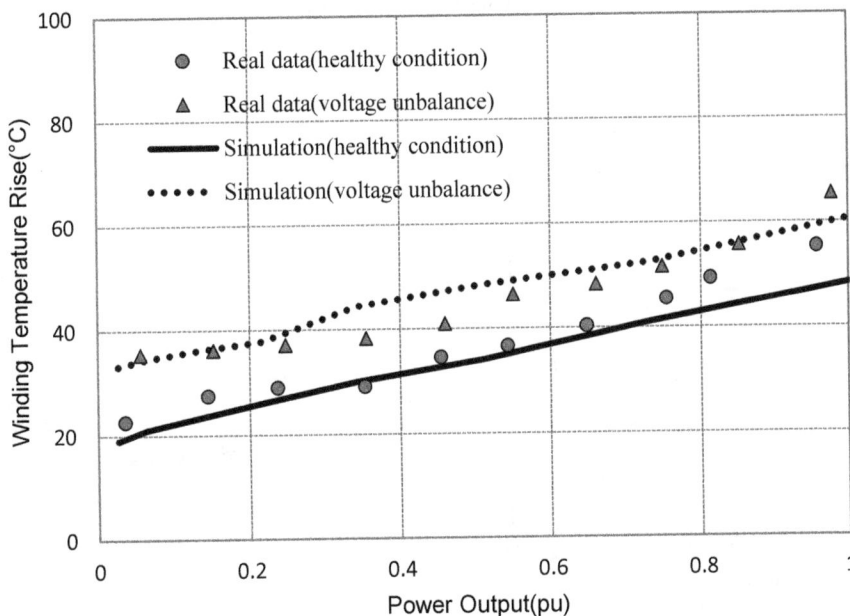

Figure 14. Winding temperature rise against normalized power output when the DFIG is under healthy conditions and with voltage unbalance.

The thermal mechanism due to voltage unbalance within the generator is different from a ventilation system failure. Voltage unbalance will have a direct impact on the generator output current, which further affects the losses in the generator. In this case, both the current and resistance of the stator winding increases while the thermal resistance and capacity is the same as the healthy situation. The result shows that a 5% overvoltage on the DFIG leads to a 6–12 K (or °C) temperature increase in the generator stator winding. The magnitudes of the temperature increases are nearly in proportion for different power generations. It is proved by the same gradient as shown in the plot. By combining both simulation and real data analysis, a curve of winding temperature against power output can reveal certain thermals mechanism in the DFIG. Failure modes related to the change of thermal behaviors such as ventilation faults and voltage unbalance can be distinguished clearly using this approach. The approach is thus proven to be useful for fault detection and diagnosis of WT DFIGs and the simulation model plays a key role in providing guidance for post-data analysis and interpretation.

3.4. Influence of Wind Speed and Failure Mode on Winding Lifetime

Assuming Class F insulation for a generator stator winding, Figure 15a shows the estimated lifetime of a WT DFIG operating at a certain wind speed. Under healthy operating conditions, the estimated lifetime of the DFIG will be reduced by an increase of wind speed. Figure 15a also shows the estimated lifetime of stator windings subjected to two failure modes, *i.e.*, voltage unbalance with 10% overvoltage and ventilation system failure with thermal resistance increasing by 12.5% and thermal capacitance decreasing by 46.2%. In the higher wind speed range above 7 m/s, ventilation failure has a greater impact on shortening generator lifetime. However, the voltage unbalance has nearly the same effects on shortening generator lifetime at different wind speeds, so the estimated lifetime curve is shifting downwards vertically. Figure 15b shows the estimated lifetimes of stator windings with different Class ratings. To choose an appropriate generator insulation Class, a tradeoff should be made between lifetime design and overall cost.

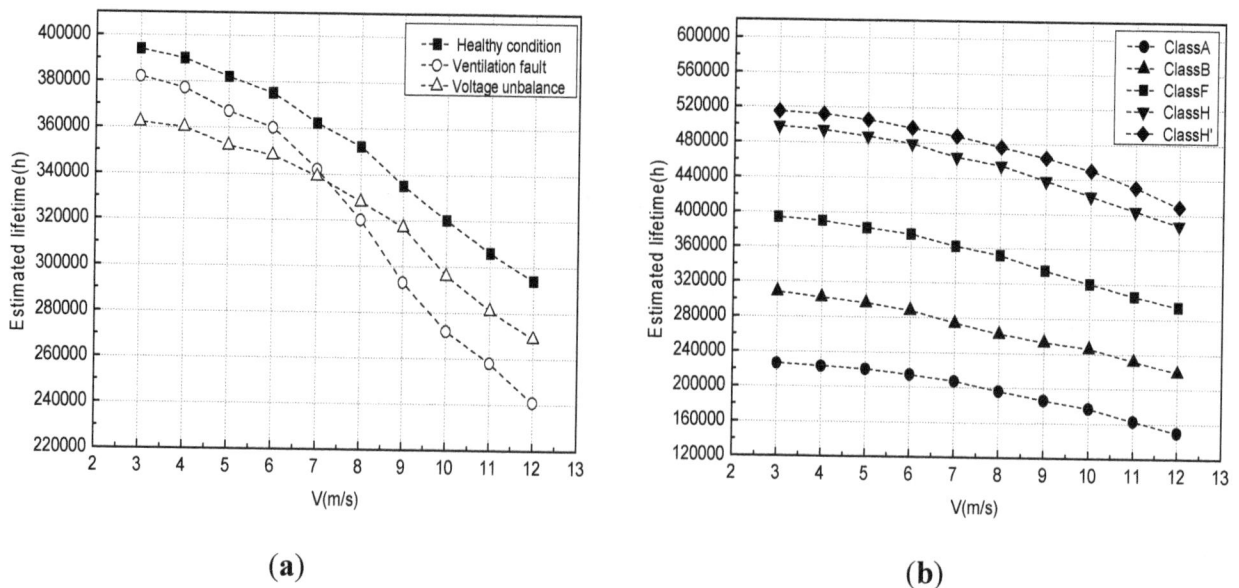

(a)   **(b)**

Figure 15. Lifetime downrating of stator windings subjected to generator system faults (a); lifetime estimation of stator windings with different Class ratings (b).

4. Conclusions

The paper developed an electro-thermal model for the DFIG in a WT. The generator power loss mechanism is analyzed and a simplified thermal network model is formed using lumped circuit parameters. Then it is incorporated into a WT system model to study the influence of wind speed on the steady and transient thermal behaviors of DFIG stator windings. The conclusions are as follows:

(1) The thermal mechanism within the WT generator is different from that of conventional machines with constant rotational speed, *i.e.*, the power losses within a WT generator are mostly affected by the wind speed profile, which further determines the magnitude of the stator winding temperature.

(2) While the long-term/steady state temperature characteristics of WT generator are determined by the power loss mechanism, the short-term/transient characteristics are determined by the thermal parameters, *i.e.*, the thermal resistance and capacitance reflect the thermal process within the generator.

(3) A generator ventilation failure will cause an increase of winding resistance, and thermal resistance of the LPN model. This essentially reflects an interaction effect between power losses and the thermal process of the ventilation system within a generator.

(4) Supply voltage unbalances will cause an increase of generator output currents and an increasing magnitude of power losses within the generator. In this process, the electro-thermal effects dominate and thermal parameters remain the same.

(5) In the curve of stator winding temperature rise against power output, depending on the specific failure mode, the gradient and intercept of the curve will be different. Simulation clearly demonstrates the physical meanings corresponding to the curve changes, based on which effective fault detection and diagnosis can be easily implemented and interpreted.

(6) Both wind speed and failure mode have negative impacts on generator lifetime. At high wind speed, ventilation system failures will cause more damage to the generator.

Acknowledgments

This is a research work funded by The Natural Science Foundation of Jiangsu Province (BK2013135), Start-up Scientific Research Project of NUST, Jiangsu Top Six Talent Summit Fund (ZBZZ-045). Returned Overseas Students Preferred Funding. The Fundamental Research Funds for the Central Universities, No. 30915011324.

Author Contributions

All the authors contributed equally to the theoretical model development, data analysis and interpretation of the comparison results.

Conflicts of Interest

The authors declare no conflict of interest.

References

1. Perveen, R.; Kishor, N.; Mohanty, S.R. Off-shore wind farm development: Present status and challenges. *Renew. Sust. Energ. Rev.* **2014**, *29*, 780–792.

2. Liserre, M.; Cardensas, R.; Molinas, M.; Rodriguez, J. Overview of Multi-MW wind turbines and wind parks. *IEEE Trans. Ind. Electron.* **2011**, *58*, 1081–1095.

3. Nienhaus, K.; Hilbert, M.; Baltes, R.; Bernet, C. Statistical and time domain signal analysis of the thermal behavior of wind turbine drive train components under dynamic operation conditions. *JPCS* **2012**, *364*, 012132.

4. Ribrant, J.; Bertling, L. Survey of failures in wind power systems with focus on Swedish wind power plants during 1997–2005. *Trans. Energy Conversion* **2007**, *22*, 167–173.

5. Inoue, A.; Takaahashi, R.; Murata, T.; Tamura, J.; Kimura, M.; Futami, M. A calculation method of the total efficiency of wind generators. *Electr. Eng. Jpn.* **2006**, *157*, 52–62.

6. Stone, G.; Boulter, E.A.; Culbert, I.; Dhirani, H. *Electrical Insulation for Rotating Machines—Design, Evaluation, Aging, Testing, and Repair*; Wiley-Interscience: Hoboken, NJ, USA, 2004; pp. 137–179.

7. Alberti, L.; Bianchi, N. A Coupled thermal–electromagnetic analysis for a rapid and accurate prediction of IM performance. *IEEE Trans. Ind. Electron.* **2008**, *55*, 3575–3582.

8. Boglietti, A.; Cavagnino, A.; Staton, D.; Shanel, M.; Mueller, M.; Mejuto, C. Evolution and modern approaches for thermal analysis of electrical machines. *IEEE Trans. Ind. Electron.* **2009**, *56*, 871–882.

9. Demetriades, G.D.; De La Parra, H.Z.; Andersson, E.; Olsson, H. A real-time thermal model of a permanent-magnetic synchronous motor. *IEEE Trans. Power Electron.* **2010**, *25*, 463–474.

10. Oraee, H. A quantative approach to estimate the life expectancy of motor insulation systems. *IEEE Trans. Dielectr. Electr. Insul.* **2000**, *7*, 790–796.

11. Takahashi, R.; Ichita, H.; Tamura, J. Efficiency calculation of wind turbine generation system with doubly-fed induction generator. In Proceedings of the 2010 XIX International Conference on Electrical Machines (ICEM), Rome, Italy, 6–8 September 2010.

12. Xu, L.; Wang, Y. Dynamic modeling and control of DFIG-Based wind turbines under unbalanced network conditions. *IEEE Trans. Power Syst.* **2007**, *22*, 314–323.

13. Brancato, E.L. Estimation of lifetime expectancies of motors. *IEEE Electr. Insul. Mag.* **1992**, *8*, 5–13.

14. Nienhaus, K.; Hilbert, M. Thermal analysis of a wind turbine generator by applying a model on real measurement data. In Proceedings of the 2012 IEEE International Workshop on Applied Measurements for Power Systems (AMPS), Aachen, Germany, 26–28 September 2012.

15. Siddique, A.; Yadava, G.S.; Singh, B. Effects of voltage unbalance on induction motors. In Proceedings of the 2004 IEEE International Symposium on Electrical Insulation, Indianapolis, IN, USA, 19–22 September 2004.

16. Pillay, P.; Manyage, M. Loss of life in induction machines operating with unbalanced supplies. *IEEE Trans. Energy Conversion* **2006**, *21*, 813–822.

Application of a Backfilling Method in Coal Mining to Realise an Ecologically Sensitive "Black Gold" Industry

Xiaowei Feng [1,2,†], Nong Zhang [1,2,3,*], Lianyuan Gong [4,†], Fei Xue [1,2] and Xigui Zheng [1,2]

[1] Key Laboratory of Deep Coal Resource Mining, Ministry of Education,
China University of Mining and Technology, Xuzhou 221116, China;
E-Mails: fxw_mining@foxmail.com (X.F.); kdckxf@126.com (F.X.); ckzxg@126.com (X.Z.)

[2] State Key Laboratory of Coal Resources and Safe Mining, School of Mines,
China University of Mining and Technology, Xuzhou 221116, China

[3] Hunan Key Laboratory of Safe Mining Techniques of Coal Mines,
Hunan University of Science and Technology, Xiangtan 411201, China

[4] Key Laboratory of Coal Processing and Efficient Utilization of Ministry of Education,
School of Chemical Engineering and Technology, China University of Mining and Technology,
Xuzhou 221116, China; E-Mail: gly_huagong@foxmail.com

[†] These authors contributed equally to this work.

[*] Author to whom correspondence should be addressed; E-Mail: zhangnong@126.com

Academic Editor: Xiangzheng Deng

Abstract: China, as the largest coal-producing and -consuming country in the world, is highly dependent on its coal industry, or "Black Gold" industry, for the national energy and economy. The consequent environmental crises, however, have persisted for decades, and the most serious effect is surface subsidence induced by underground mining. Underground coal excavation in China has ignored this problem for thousands of years, even though it causes conspicuous damage to the surface ecosystem and construction projects due to the subsidence of overlying strata. This study recommends paste backfilling to replace the space originally occupied by coal resources to avoid such subsidence and proposes backfilling schemes for two mainstream mining methods used in China's collieries, namely, continuous mining and fully mechanised coal mining. These methodologies have been successfully implemented in some collieries, and the gob area can be backfilled immediately to prevent surface subsidence. To promote an ecological ideology when conflict exists between economic profits and environmental protection, experience from developed countries should be considered, support and appropriate legislation from the government are essential, and the perspective of colliery managers should be taken into account, and further in-depth study on strata subsidence and backfilling material must be pursued.

Keywords: underground mining; surface subsidence; "Black Gold" industry; backfilling method; environmental protection; ecological sensitivity

---

1. Introduction

Worldwide, 30.1% of primary energy, 40% of electricity generated, and 70% of steel manufacturing depend on coal, and the total global coal production in 2013 was 7,822.8 Mt. Regional contributions to this total are available from the International Energy Agency and the BP Statistical Review of World Energy [1,2]. The data show that China was the largest coal producer (45.6%) in 2013; thus, China continues to play a leading role in environmental disruption even though the nation resolved to reduce the carbon intensity of its economy (the amount of CO_2 emitted per unit of GDP) by 40%–45% during 2005–2020 [3]. It does not make sense to place greater value on production or economic development than on the protection of our environment, yet the environmental crises resulting from both coal mining and use, including soil erosion, water pollution, surface subsidence, the release of pollutants such as oxides of sulphur (SO_x) and nitrogen (NO_x), and impacts on local biodiversity, persist.

These crises comprise a number of environmental challenges; one such challenge that urgently needs to be addressed is surface subsidence [4]. Many factors may lead to surface subsidence, such as the collapse of columns underground or invisible karst rivers among strata. The most prominent factor causing subsidence is underground mining [5,6]; when a block of sufficient width and length is excavated from a coal field, the original equilibrium state is broken up regardless of whether the mining method used is longwall mining or room and pillar mining. Consequently, rock strata between the immediate roof of the coal seam and the ground surface can be moved. During our research, we investigated coal mine fields in Xuzhou City, where the ground surface has been particularly damaged by underground mining, as shown in Figure 1. It is clear that subsidence around the coal mine field has become a serious concern, and unlike other issues, such subsidence is not only a technical problem but also a socially relevant issue.

Many studies have been carried out around the world to evaluate surface subsidence for the purpose of developing corresponding countermeasures [7–11], but what if we change our approach from the outset, taking measures to prevent this type of subsidence from occurring? A potential solution to these questions may lie in radically preventing subsidence induced by mining activity at its source, *i.e.*, by using some other material to fill the space occupied by the extracted coal. Traditional schemes include waste rock backfilling [12], coal ash backfilling, and paste backfilling, among others. With measures such as these, ground surface subsidence can be reduced to some extent. However, backfilling is not mandatory in China's collieries as this environmentally sound practice has long been perceived as a costly, time-consuming, and low-efficiency engineering method.

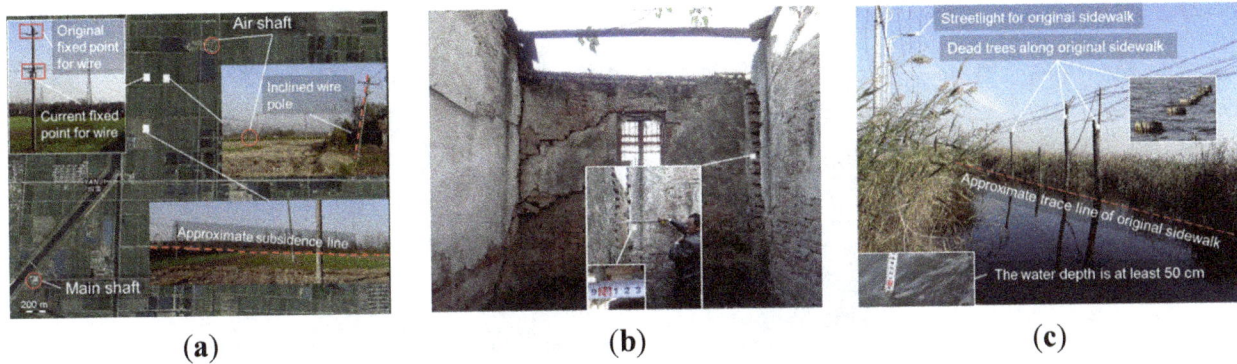

Figure 1. Example of surface subsidence problems induced by underground mining. (**a**) Surface subsidence around Zhangji Colliery in Xuzhou, at latitude and longitude of 34.3667 and 117.0229, respectively; the mining depth is more than 900 metres; (**b**) Building damage associated with Zhangji Colliery in Xuzhou, at latitude and longitude of 34.3846 and 117.0315, respectively; (**c**) Waterlogging around the Pangzhuang Colliery in Xuzhou, at latitude and longitude of 34.3566 and 117.0766, respectively. (Xiaowei Feng and Lianyuan Gong performed investigations around the subsidence areas and obtained these figures by themselves).

Hence, the status since the 1970s has remained the use of backfilling. Currently, a simple, natural caving method in which the roof above the coal seam is allowed to naturally cave in under the effects of gravity is mainly adopted because it incurs no extra cost; this method can easily lead to surface subsidence. It is critical that innovative backfilling methods be developed to create a more ecologically sound "Black Gold" industry, and such methods must provide both economic and technical value.

2. Characteristics of Surface Subsidence Induced by Underground Mining

As a result of continuous sedimentation effects and tectonic movements occurring on the scale of billions of years, the Earth's surface shows a general trend toward stability, even though seismic activities and volcanism are spread worldwide. However, within the Earth, artificial underground mining poses enormous hazards to the safety of miners and the environment. Surface subsidence induced by mining has long been ignored because governments focus more on financial gains than environmental protection or simply because most people are more familiar with greenhouse gas emissions and naturally consider emissions to be the greatest problem caused by underground coal excavation.

The extraction of underground coal resources has been undertaken for thousands of years in China, according to the record in *The Tiangong Kaiwu*, or *The Exploitation of the Works of Nature*, a Chinese encyclopaedia compiled by Song Yingxing, initially published in May 1637 (see Figure 2a). Once excavation occurs within the ground, the overlying rock strata experience subsidence, as illustrated in Figure 2b. It is clear that the associated damage compensation is much more expensive than the economic profits from excavation. Lakes, rivers, reservoirs, and other water bodies are easily disturbed, causing mine flooding underground, while the surface is subjected to water erosion/pollution, desertification, and building damage.

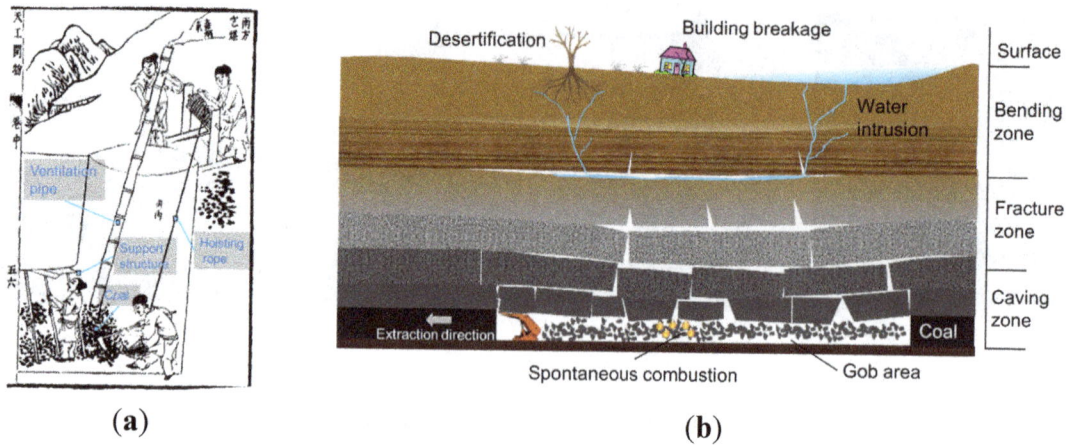

Figure 2. Comparison of an ancient mining method and current mechanised mining without backfilling. (**a**) The ancient mining method described in *The Tiangong Kaiwu*, published in 1637; (**b**) The current mechanised mining method; the method is characterised by its caving system, which is both economical and efficient to enable high productivity.

Figure 2b also indicates an apparent three-zone division along the vertical direction from the mining level to the surface level—the caving zone, fracture zone, and bending zone. This zone division is based on the degree of fracture of strata overlying the coal seam, and the degree of fracture is in turn associated with the mining method, spatial relationships, geotechnical conditions of the rock strata, and excavation duration [13]. It is generally thought that these three zones can only develop when the ratio of the thickness of overlying strata to the thickness of the coal seam is greater than 40, but this ratio is surpassed extremely easily in the case of the many 1,000-metre-deep collieries; data show that there are already more than 50 collieries with an excavation depth greater than 1,000 metres in China [14]. Underground fracturing induces varying hazards on the surface, such as water loss and soil erosion, building damage, and ecological disruption, among others. Moreover, the insufficient backfilling produced by single roof caving behind hydraulic supports is inadequate to prevent the deformation of the overlying rock mass. Some unfilled space can remain in the gob area, which causes problems for the excavation of the neighbouring working face because the dead weight of the surrounding rock mass is only partially supported by the caving gangue in the gob area, and the remaining weight must be transferred to the adjacent intact coal body to maintain stability. In other words, this transfer causes a pressure increase on the neighbouring working face, thus increasing the costs required to maintain stability in the adjacent face. Additionally, it is impossible to realise a one hundred percent of coal recovery ratio, which means certain amount of coal can be left over in gob area, then spontaneous combustion may be incurred under the influence of unstable environment and oxygen in the gob area, which presents a danger to both the efficiency and safety of production [15]. Surface ecology and construction are also disturbed by subsidence. Such damage is especially predominant in the north-western coal districts of China, like Qinghai Province, Sinkiang Province, Ningxia Province, *etc.*, where the ecological environment, primarily semi-desert grassland, is extremely fragile. Thus, disturbances underground readily lead to surface water losses and can cause environmental conditions to turn from bad to worse.

Injecting grout into the bed separation in the overburden of coal seams may serve as a solution to this environmental problem [16]. Such an approach is a remedial measure, though it has long been treated as an effective measure to seal off spaces between strata to resist on-going subsidence, it cannot rehabilitate the ground surface to its original landform. Furthermore, injecting grout can only take place when bed separation does occur between strata, and the costs and technical problems can be much greater than an immediate backfilling method following coal mining. Immediate backfilling is conducted from the outset to backfill the space originally occupied by coal resources (see Figure 3). This method is a win-win situation from both an economic and ecological perspective.

Figure 3. Current mechanised mining with backfilling procedure employed (drawing designed by Xiaowei Feng).

Figure 3 demonstrates that timely backfilling prevents the occurrence of the three-zone division above the coal seam, and the ecosystem on the ground is also minimally influenced. In conclusion, mining that protects water resources and is environmentally friendly, or green mining, is a challenge, and it is urgent that we turn this previously economically driven engineering practice into an ecological engineering practice.

3. Methodology for Surface Subsidence Prevention

As mentioned in the previous sections, the key to preventing surface subsidence lies in the timely filling of man-made spaces underground with alternative materials after excavation. The alternatives can vary from waste rock backfilling [17], coal ash backfilling and hydraulic sand backfilling to paste backfilling and similar measures. Gob backfilling makes it possible to excavate coal resources trapped under artificial constructions and natural water bodies [18], and this method is thus increasingly used in underground mines all over the world [19]. Currently, there are nearly 13.79 billion tons of coal trapped under the aforementioned conditions in key state-owned collieries in China [20]. To choose the most effective of these alternatives, the following equation can be utilised during practical field trials:

$$\frac{\rho}{\rho'} = k \frac{M}{M'} \tag{1}$$

where
M is weight of coal excavated, in tons.
M' is weight of backfilling material to be applied, in tons.
ρ is the density of coal, in tons per cubic metre (1.3 t/m^3 for coal, in general).
ρ' is the density of backfilling material, in tons per cubic metre (1.6 t/m^3 for backfilling material, in general).
k is a backfilling factor (0.6–0.95) that varies by system; the general value for coal mines is 0.8.

Based on the above specific values, it can be found that $M'/M \approx 1$. It is clear that the amount of backfilling material required is approximately equal to the coal production. It is difficult to prepare such a large amount of material and transfer it to the space behind the coal face, especially for measures like waste rock backfilling, coal ash backfilling, and hydraulic sand backfilling as a result of their very one-dimensional technical system. This is the greatest disadvantage to backfilling because of the dramatic increase in cost, which may be twice the production profits from coal excavation. However, among these alternatives, the most effective method is paste backfilling [21–23] by mixing waste rock, aeolian sand, or tailings with coal ash and cement. First, a certain amount of chemical admixtures and auxiliary material is added to upgrade the pumpability of the mixture, and then the mixture is pumped to the rear space of the coal excavation line. This approach has the potential to reduce surface subsidence and is a promising method for large amounts of waste reuse [24]. The aforementioned technical procedure is illustrated in Figure 4.

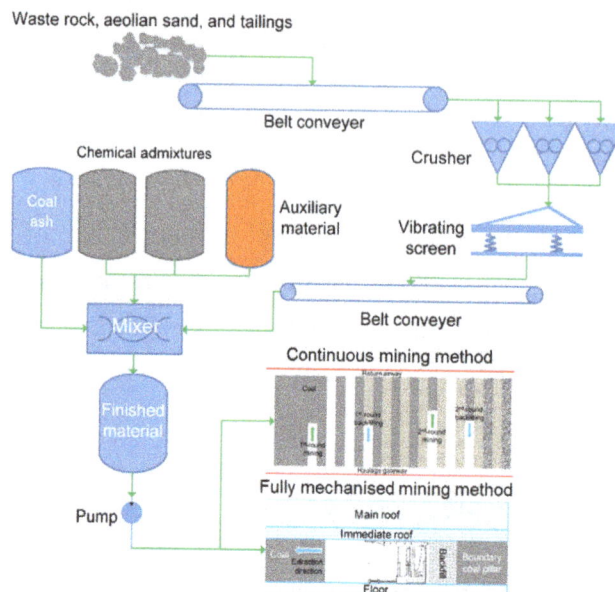

Figure 4. Diagram of underground paste backfilling (drawing designed by Xiaowei Feng).

Generally, the objectives of paste backfilling should be as follows: to geotechnically ensure a safe and efficient mining operation; to increase the extraction ratio and decrease water consumption; to decrease soil contamination and surface subsidence; and most importantly, to decrease cost compared with other measures and let the value of M'/M approach 1 as closely as possible because of the backfill material's expansion ability.

The pumpability of backfill material will dramatically decrease the cost of transportation. Thus, the pumping pressure should be lowered and the hydration time should be increased to allow for longer transportation distances. The backfilling material should possess appropriate expansion and solidification properties to decrease porosity and improve uniformity throughout the gob area when it reaches the excavated site. The material can thus support the roof and prevent the subsidence of overlying strata [25]. Two mainstream methodologies are used: the first addresses continuous mining faces, and the second is designed for fully mechanised coal mining faces. The schematic details for backfilling methodology in a continuous mining face are shown in Figure 5, the coal seam is divided into strips by a skip-mining method. The first round of mining leaves a certain width of coal pillars, this preliminary coal excavation is a type of room and pillar method, where pillars are left to support the roof in case of roof collapse or further subsidence of the overlying strata and the coal is excavated for economic profit. At a certain distance from the first-round mining face, the first-round backfilling face is arranged to fill up the space initially occupied by coal mass. These two faces are capable of excavating a certain amount of coal while simultaneously controlling the potential subsidence of the overlying roof. However, nearly half of the coal is left behind, and it is thus proposed to apply a second-round mining face to excavate the remaining coal pillars among the previous first-round backfilling faces. The same volumes of gob area occur when excavating these remaining coal pillars; hence, to thoroughly prevent roof subsidence, a second-round backfilling face behind the second-round mining face must completely fill the gob area. In this type of backfilling method, these four operational sequences are relatively independent. Furthermore, this method both guarantees a high recovery rate of coal and controls surface subsidence. Such backfilling has been successfully employed in the Yuyang colliery in Shanxi Province, where the total subsidence area before backfilling was 4.6398 square kilometres on the ground [26]. Another successful case is the Xuchang Colliery in the city of Zibo in Shandong Province [27].

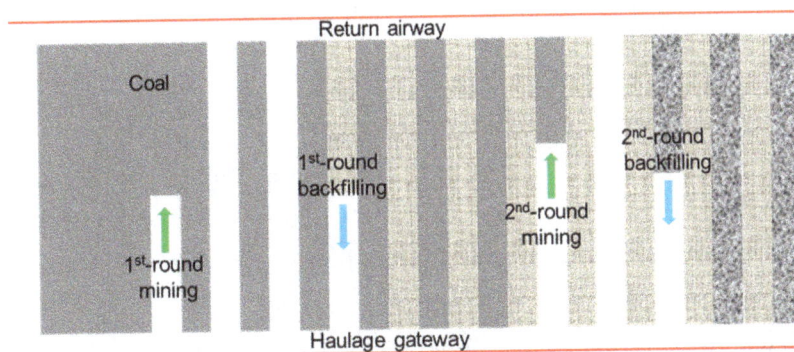

Figure 5. Schematic illustration of backfilling methodology in a continuous mining face (drawing designed by Xiaowei Feng).

For fully mechanised coal mining, a different backfilling method should adopted to address the unique nature of the coal mining face and equipment used. The hydraulic support is characterised by its backfilling function, which is capable of pumping paste backfilling material into the gob area behind the support. The working procedure can be divided into four steps (see Figure 6). The movement of the support along the extraction direction allows the mould bag to stretch automatically (see Figure 6b) thus allowing backfilling material to be pumped in (see Figure 6c). One cycle is finished when the previously pumped material coagulates, and the process is repeated throughout the coal mining process, as shown in Figure 6d. Apparently, this approach can also fully pack the gob area and confine the subsidence of overlying roof strata such as the immediate roof and main roof, thus supporting the concept of "green mining" [28]. The immediate roof generally consists of siltstone, sandy shale, and shale, which can remain stable for a certain period even if the coal seam underneath it has been mined out for some distance, but will eventually sag due to the gap below if no backfilling procedure has been applied. The main roof is much stiffer and thicker than the immediate roof, which will apply greater pressure on the mining face, underlying strata, and related galleries below;

this relationship also highlights the importance of utilising backfilling strategies during mining activities. This type of backfilling technology is also widely being used at the Zhucun Colliery in the city of Jiaozuo in Henan Province [29].

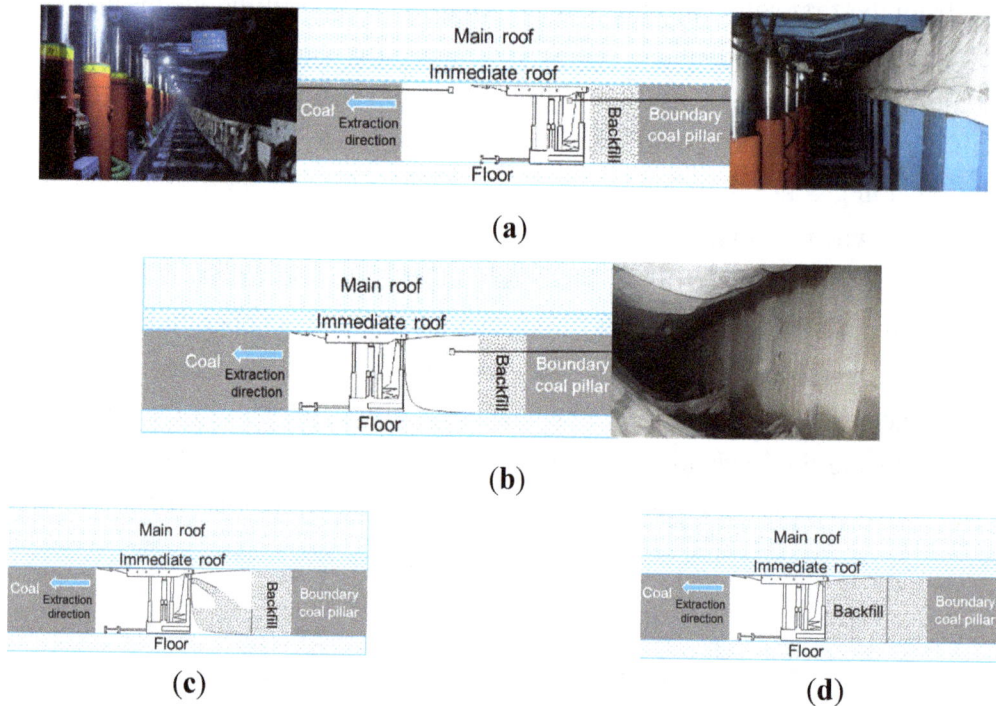

(a)

(b)

(c)

(d)

Figure 6. Backfilling procedure for fully mechanised coal mining with self-backfilling hydraulic supports. **(a)** The moment when the first-round backfilling neighbouring the boundary coal pillar is finished; the left and right photos, taken in the Yulin colliery of Shanxi Province, indicate corresponding sections in the centre sketch; **(b)** The left sketch shows the moment when the mould bag stretches as a result of the forward movement of the hydraulic support, which is also indicated in the right photo, also taken in the Yulin colliery of Shanxi Province; **(c)** Backfilling material being pumped into the rear space formed by the mould bag; **(d)** End of the first-round backfilling procedure. (drawings in this figure were designed by Xiaowei Feng and photos were taken by Xigui Zheng and Fei Xue in a related field trial in Shanxi Province).

4. Discussion and Future Outlook

National legislation must support both economic development and ecological sustainability. Currently, China is still a developing country with an energy economy that is largely dependent on its coal resources. Although China's annual coal production accounts for a large part of the global total, the production method applied in collieries still lags behind those used in developed countries such as the United States, Germany, and Australia. Many workers contribute a lifetime of labour, or even their lives, to the national energy industry. Legislation under these circumstances cannot limit the widespread effects of collieries in a short time, which may even cause a hysteresis phenomenon in technical development in coal production. However, legislation addressing the environmental crises induced by underground mining is imperative, and during this process, China can benefit from other countries' experiences. The Ruhr District in Germany can serve as an example, where land subsidence was extremely serious due to large-scale coal extraction over the past 100 years [30], yet the regional initiatives implemented by the government since the 1970s have turned this district into a world-renowned "Garden Industrial Zone" in view of its current diversification of economy, emerging industry, environment rehabilitation, and other factors. As another example, consider the United States, where the conflict between environmental protection and growing coal excavation has been on-going since the 1970s. In the Yellowstone River Basin of eastern Montana, the abundant, high-quality water and the nation's largest coal reserves presented a great conflict of interest because the extraction of the latter requires large volumes of water [31]. In the late 1970s, south-eastern Montana's Tongue River basin was experiencing rapid development of its extensive coal deposits, which was greatly impacting the basin's hydrologic systems, and the Water Use Act required certain adjustments to relieve this situation despite some unexpected restrictions [32]. Under such a historical background, US president Jimmy Carter signed the Surface Mining Control and Reclamation Act (SMCRA) in 1977. SMCRA is the first mining environmental protection law in the USA, and it provides standards for environmental protection related to mining activity. Simultaneously, the Office of Surface Mining (OSM) was set up to supervise the protection and rehabilitation of environment [33].

Obviously, experience from these countries should be examined and adjusted based on the unique mining conditions in China. A scientifically feasible and ecologically friendly coal mining industry is a key element for constructing green homes in mine fields, though this notion has only gained attention in recent years in China. Whether the coal mining industry, or "Black Gold" industry, can be realised ecologically depends on both the rationality of the economic structure and the feasibility of its technological background. The government should consider that coal is the main resource for and an extremely reliable source of energy in China, but coal represents a raw product in the national industrial chain and the extent of the nation's reliance on this industry is still unreasonable. Hence, green mining technology deserves support and concern from the government, which will undoubtedly urge supervisors of collieries to place more importance on the environment while simultaneously avoiding negative influences on the economy. From a technological perspective, more research and investments should be focused on basic study of strata subsidence arising from underlying coal mass excavation, including cost optimisation of the paste backfilling process and related backfilling materials. Moreover, different collieries have different characteristics; it should be emphasised that collieries in eastern China should focus more on how to protect the surface ground and construction, whereas those in western China should focus more on how to protect their fragile ecosystems and barren water system. Ecosystem protection patterns and economic evaluation systems should also be studied, including their relationship with enterprise cost, which can thus guide government legislation.

5. Conclusions

The Chinese economic structure is largely dependent on its coal production, which causes inevitable problems to our shared ecosystem. This paper focuses on the surface subsidence caused by underground mining and proposes corresponding possible solutions, which can be generalised as follows:

(1) Underground mining without additional backfilling has an enormous impact on the overlying strata above coal seams. The direct consequence to the surface is subsidence, which then triggers building damage, water resource erosion or pollution, and desertification.

(2) Among alternatives that may mitigate surface subsidence, this paper recommends backfilling with paste, which is characterised by its pumpability, convenient underground transportation, and high expansion as well as low porosity in the gob area. Paste backfilling thus has the multiple advantages of low cost, high efficiency, and technical feasibility compared with other measures.

(3) In view of the two most commonly used mining methods in Chinese collieries, *i.e.*, fully mechanised coal mining and continuous mining, this paper suggests different backfilling processes for each method. Both approaches are capable of stabilising the overlying strata and mitigating surface subsidence due to underground mining.

(4) As to the future of China's coal industrial development and environmental protection, legislation is still imperative and should consider both successful experiences in developed countries and the current unique situations in China. The key to success lies in finding an equilibrium point between government officers and colliery managers. Additional studies pertinent to these issues should also be carried out to transform this environmentally damaging industry into a green ecological "Black Gold" industry.

Acknowledgments

Financial support for this paper from the Fundamental Research Funds for the Central Universities (NO. 2014XT01), Program for Changjiang Scholars and Innovative Research Team in University (IRT_14R55), and National Basic Research Program of China (2013CB227904) is gratefully acknowledged.

Author Contributions

X.F. and X.Z. proposed the idea. N.Z. provided suggestions and field trials in Shanxi Province. X.F. and L.G. performed investigations around subsidence areas in the city of Xuzhou in Jiangsu Province. X.F. wrote and revised the manuscript. F.X. assisted X.F. in revising the manuscript. X.F. created the Figures. N.Z. supervised the project.

Conflicts of Interest

The authors declare no conflict of interest.

References

1. Key World Energy Statistics 2014. Available online: http://www.iea.org/publications/freepublications/publication/key-world-energy-statistics-2014.html (accessed on 20 March 2015).
2. Statistical Review of World Energy 2014. Available online: http://www.bp.com/en/global/corporate/about-bp/energy-economics/statistical-review-of-world-energy.html (accessed on 20 March 2015).
3. Guan, D.B.; Klasen, S.; Hubacek, K.; Feng, K.S.; Liu, Z.; He, K.B.; Geng, Y.; Zhang, Q. Determinants of stagnating carbon intensity in China. *Nat. Clim. Change* **2014**, *4*, 1017–1023.

4. Cui, F.; Liu, P.L.; Feng, R.M.; Zhang, H.X. Coal mining present application status of coal backfill in China. In Proceedings of the 30th Annual International Pittsburgh Coal Conference 2013, PCC 2013, Beijing, China, 15–18 September 2013; pp. 616–628.

5. Hu, Z.Q.; Yang, G.H.; Xiao, W.; Li, J.; Yang, Y.Q.; Yu, Y. Farmland damage and its impact on the overlapped areas of cropland and coal resources in the eastern plains of China. *Resour. Conserv. Recy.* **2014**, *86*, 1–8.

6. Bell, F.G.; Stacey, T.R.; Genske, D.D. Mining subsidence and its effect on the environment: Some differing examples. *Environ. Geol.* **2000**, *40*, 135–152.

7. Jung, Y.B.; Song, W.K.; Cheon, D.S.; Lee, D.K.; Park, J.Y. Simple method for the identification of subsidence susceptibility above underground coal mines in Korea. *Eng. Geol.* **2014**, *178*, 121–131.

8. Sheorey, P.R.; Loui, J.P.; Singh, K.B.; Singh, S.K. Ground subsidence observations and a modified influence function method for complete subsidence prediction. *Int. J. Rock Mech. Min.* **2000**, *37*, 801–818.

9. Kim, K.D.; Lee, S.; Oh, H.J.; Choi, J.K.; Won, J.S. Assessment of ground subsidence hazard near an abandoned underground coal mine using GIS. *Environ. Geol.* **2006**, *50*, 1183–1191.

10. Zhao, C.Y.; Lu, Z.; Zhang, Q.; Yang, C.S.; Zhu, W. Mining collapse monitoring with SAR imagery data: A case study of datong mine, China. *J. Appl. Remote Sens.* **2014**, *8*, 083574.

11. Blachowski, J.; Chrzanowski, A.; Szostak-Chrzanowski, A. Application of GIS methods in assessing effects of mining activity on surface infrastructure. *Arch. Min. Sci.* **2014**, *59*, 307–321.

12. Zhang, Q.; Zhang, J.X.; Zhao, X.; Liu, Z.; Huang, Y.L. Industrial tests of waste rock direct backfilling underground in fully mechanized coal mining face. *Environ. Eng. Manag. J.* **2014**, *13*, 1291–1297.

13. Lian, X.G.; Jarosz, A.; Saavedra-Rosas, J.; Dai, H.Y. Extending dynamic models of mining subsidence. *Trans. Nonferrous Met. Soc.* **2011**, *21*, 536–542.

14. China National Coal Association (CNCA); Shandong Energy Group Co., Ltd. *Mining Technology for Thousand-Metre-Deep Pits around Countrywide Collieries*; China University of Mining and Technology Press: Xuzhou, Jiangsu, China, 2013; pp. 2–23.

15. Zhou, C.C.; Liu, G.J.; Cheng, S.W.; Fang, T.; Lam, P.K.S. The environmental geochemistry of trace elements and naturally radionuclides in a coal gangue brick-making plant. *Sci. Rep.* **2014**, *4*, 6221.

16. Xuan, D.Y.; Xu, J.L. Grout injection into bed separation to control surface subsidence during longwall mining under villages: Case study of Liudian coal mine, China. *Nat. Hazards* **2014**, *73*, 883–906.

17. Zhang, J.X.; Miao, X.X.; Guo, G.L. Development status of backfilling technology using raw waste in coal mining. *Int. Min. Saf. Eng.* **2009**, *26*, 395–401.

18. Fan, G.W.; Zhang, D.S.; Wang, X.F. Reduction and utilization of coal mine waste rock in China: A case study in Tiefa coalfield. *Resour. Conserv. Recy.* **2014**, *83*, 24–33.

19. Li, L. A new concept of backfill design-Application of wick drains in backfilled stopes. *Int. J. Min. Sci. Technol.* **2013**, *23*, 763–770.

20. Zhang, J.X.; Zhou, N.; Huang, Y.L.; Zhang, Q. Impact law of the bulk ratio of backfilling body to overlying strata movement in fully mechanized backfilling mining. *J. Min. Sci.* **2011**, *47*, 73–84.

21. Mez, W.; Schauenburg, W. Backfilling of caved-in goafs with pastes for disposal of residues. In Proceedings of the 6th International Symposium on Mining with Backfill, Brisbane, Australia, 14–16 April 1998; Volume 98, pp. 245–248.

22. Wang, X.M.; Zhao, J.W.; Xue, J.H.; Yu, G.F. Features of pipe transportation of paste-like backfilling in deep mine. *J. Cent. South Univ. Technol.* **2011**, *18*, 1413–1417.

23. Zhang, X.G.; Jiang, N.; Shang, X.L. Phy-chemical properties experiment research on coal mine paste backfilling. In Proceedings of the 3rd International Workshop on Mine Hazards Prevention and Control, Brisbane, Australia, 19–21 November 2013; Guo, W., Shen, B., Tan, Y., Cheng, W., Yan, S., Eds.; 2013; Volume 94, pp. 531–538.

24. Bian, Z.F.; Miao, X.X.; Lei, S.G.; Chen, S.E.; Wang, W.F.; Struthers, S. The challenges of reusing mining and mineral-processing wastes. *Science* **2012**, *337*, 702–703.

25. Li, L.; Aubertin, M. An improved method to assess the required strength of cemented backfill in underground stopes with an open face. *Int. J. Min. Sci. Technol.* **2014**, *24*, 549–558.

26. Ma, J.G. Current situation of environment and rehabilitation research on Yuyang colliery in Shanbei. *Land Resour. Infor.* **2013**, *9*, 39–44.

27. Zhang, J.X.; Miao, X.X.; Mao, X.B.; Chen, Z.W. Research on waste substitution extraction of strip extraction coal-pillar mining. *Chin. J. Rock Mech. Eng.* **2007**, *26*, 2687–2693.

28. Liu, Z.; Zhou, N.; Zhang, J.X. Random gravel model and particle flow based numerical biaxial test of solid backfill materials. *Int. J. Min. Sci. Technol.* **2013**, *23*, 463–467.

29. Sun, X.G.; Zhou, H.Q.; Wang, G.W. Digital simulation of strata control by solid waste paste-like body for backfilling. *J. Min. Saf. Eng.* **2007**, *24*, 117–121.

30. Harnischmacher, S.; Zepp, H. Mining and its impact on the earth surface in the Ruhr District (Germany). *Z. Geomorphol.* **2014**, *58*, 3–22.

31. Thomas, J.L.; Anderson, R.L. Water-energy conflicts in Montana's Yellowstone River Basin. *J. Am. Water Resour. Assoc.* **1976**, *12*, 829–842.

32. Hickcox, D.H. Water rights, allocation, and conflicts in the Tongue River Basin, Southeastern Montana. *J. Am. Water Resour. Assoc.* **1980**, *16*, 797–803.

33. Plotkin, S.E.; Gold, H.; White, I.L. Water and energy in the Western coal lands. *J. Am. Water Resour. Assoc.* **1979**, *15*, 94–107.

4

Effects of Design/Operating Parameters and Physical Properties on Slag Thickness and Heat Transfer during Coal Gasification

Insoo Ye, Junho Oh and Changkook Ryu *

School of Mechanical Engineering, Sungkyunkwan University, Suwon 440-746, Korea;
E-Mails: mercury11@skku.edu (I.Y.); austin87@skku.edu (J.O.)

* Author to whom correspondence should be addressed; E-Mail: cryu@me.skku.ac.kr

Academic Editor: Mehrdad Massoudi

Abstract: The behaviors of the slag layers formed by the deposition of molten ash onto the wall are important for the operation of entrained coal gasifiers. In this study, the effects of design/operation parameters and slag properties on the slag behaviors were assessed in a commercial coal gasifier using numerical modeling. The parameters influenced the slag behaviors through mechanisms interrelated to the heat transfer, temperature, velocity, and viscosity of the slag layers. The velocity profile of the liquid slag was less sensitive to the variations in the parameters. Therefore, the change in the liquid slag thickness was typically smaller than that of the solid slag. The gas temperature was the most influential factor, because of its dominant effect on the radiative heat transfer to the slag layer. The solid slag thickness exponentially increased with higher gas temperatures. The influence of the ash deposition rate was diminished by the high-velocity region developed near the liquid slag surface. The slag viscosity significantly influenced the solid slag thickness through the corresponding changes in the critical temperature and the temperature gradient (heat flux). For the bottom cone of the gasifier, steeper angles were favorable in reducing the thickness of the slag layers.

Keywords: ash deposition; coal gasification; heat transfer; slag flow; syngas

1. Introduction

Many commercial coal gasification processes such as the Shell, Prenflo, Mitsubishi, Tsinghua and Siemens ones utilize entrained flow reactors feeding pulverized coal in dry or slurry forms [1–3]. Because the syngas temperature typically increases above the ash melting point, most of the coal ash is molten and deposits onto the gasifier wall to form slag layers. The fraction of slag contacting the cold wall is immobilized (solid layer), whereas that facing the hot syngas flows downward by gravity (liquid slag). At the bottom of the gasifier, the liquid slag is discharged through a slag tap to a water bath and quenched. These slag layers play an essential role in protecting the wall from excessive heat and chemical attack by the hot acidic gases.

Because of its importance in ash discharge, wall protection, and heat recovery, controlling the slag behavior is a crucial issue for the design and operation of entrained flow gasifiers [3,4]. The ash fusion temperature and slag viscosity are important properties in relation to the flowability and thickness of the liquid slag, which are determined by inherent properties such as the ash composition and chemistry, as well as external conditions such as the temperature and reaction atmosphere [5–9]. In order to minimize slag-related problems, coals with ash content and properties within an appropriate range should be selected [10,11]. For coals with high or low slag viscosity, additional supply of CaO-based flux or blending with other coals can be considered [12–15]. Together with the inherent slag properties and the design/operation parameters influence the slag behavior. For example, the syngas temperature within the gasifier dominates the depositing particle temperature and temperature profile developed within the liquid slag. Because the slag viscosity is strongly dependent on the temperature, the gas temperature can directly influence the slag thickness [16–19]. The size and shape of a gasifier are also important parameters that affect the ash deposition rate and angle of the gravity force applied to the slag [20]. However, it is not straightforward to understand their influences on the thickness, flow, and heat transfer characteristics of the slag layers, because complex mechanisms and variables are involved in the slag behaviors.

Our previous study [21] proposed a new numerical model to predict the flow and heat transfer of the slag layers, which can overcome the limitations of the analytical models of Seggani [22] and Yong et al. [20]. This model solves the governing equations for the slag layers by using the finite volume method to predict the details of their thickness, temperature, and velocity distribution along the gasifier wall. Because the model is not based on simplifications of the temperature profile and viscosity within the liquid slag layer, it can be used for various operating conditions of a gasifier and for different physical models of slag properties.

In this study, the numerical model was applied to understand the influences of the design/operation parameters and the slag properties in a large-scale coal gasifier. The parameters included the syngas temperature, mass rate and temperature of ash deposition, and height of the bottom cone in the gasifier. The slag properties included the viscosity, thermal conductivity and emissivity. These parameters were varied by 10% from the reference condition to evaluate their effects on the slag layer thicknesses, heat transfer rates and temperatures at the interfaces. The underlying physical mechanisms of the slag behaviors were examined by analyzing the details of the velocity and temperature profiles in the slag layers.

2. Numerical Methods and Test Parameters

Figure 1 shows a schematic of the Prenflo coal gasifier [22] considered in the parametric analysis. It consists of the main body, top cone and bottom cone. The bottom cone has a height of 0.30 m with a wall angle of 12°, leading to a slag tap with a radius of approximately 0.43 m.

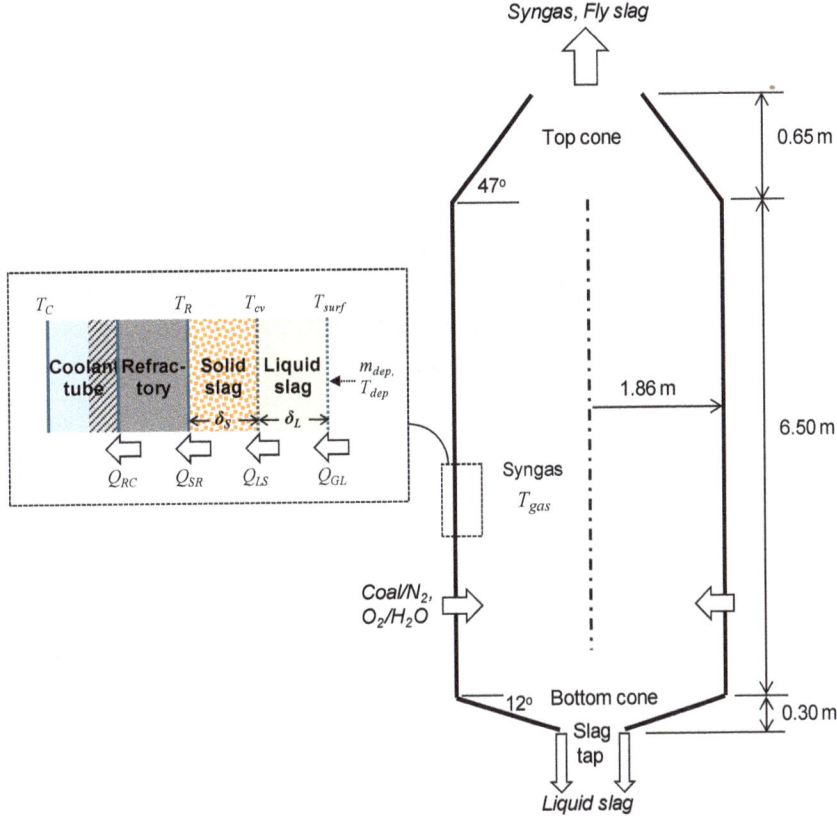

Figure 1. Schematic of the gasifier considered in this study.

A numerical model for the slag flow and heat transfer in the gasifier was presented in detail in our previous study [21]. In brief, this model considers the following sets of governing equations for the mass (m), momentum (M) and energy (H) for a control volume within the liquid slag flow:

$$m_{out} = m_{in} + m_{dep} \tag{1}$$

$$M_{out} = M_{in} + M_{dep} + \left(2\pi(r+dr)dy \cdot \mu \frac{dv}{dr}\bigg|_{r+dr} - 2\pi r dy \cdot \mu \frac{dv}{dr}\bigg|_{r} \right) + \rho g \sin\alpha \cdot dV \tag{2}$$

$$H_{out} = H_{in} + \Delta H_{react} + Q_{cond} + H_{dep} + Q_{GL} \tag{3}$$

In the above equations, the terms associated with the slag deposition (m_{dep}, M_{dep}, and H_{dep}) and heat transfer from the gas (Q_{GL}) became zero for inner control volumes. The equations are discretized using the finite volume method on the cylindrical coordinates for a section of the liquid slag layer perpendicular to the wall, and solved to determine the thickness, velocity and temperature distribution. The sum of the thicknesses of the individual control volumes becomes the liquid slag thickness (δ_L).

The solution marches from the top to the bottom of the gasifier wall in the streamwise direction. The deposition of ash onto the surface of the liquid layer along the gasifier wall is treated as the addition of a new control volume. At the other end, the solid slag layer is considered as a boundary with a no-slip condition with a fixed temperature of T_{cv}. T_{cv} represents the temperature for the critical viscosity of 25 Pa·s, below which the slag flow becomes stagnant by a rapid increase of viscosity. Under the steady-state condition, the heat transfer at the interface of the liquid and solid slag (Q_{LS}) can be used to determine the solid slag thickness (δ_S) and the interface temperatures facing the refractory (T_R). From the Fourier's law for a cylindrical system:

$$Q_{LS} = Q_{SR} = A_R k_S \frac{T_{cv} - T_R}{r_R \ln(r_R / r_S)} \tag{4}$$

$$\delta_S = r_R - r_S \tag{5}$$

Note that Q_{cond} at $r = 0$ in Equation (3) is expressed as Q_{LS} in Equation (4).

Table 1 lists the parameters associated with the gasifier design/operation and the slag properties evaluated by varying them 10% from the reference values. The reference values are based on the operating conditions in Seggiani's study [22], but were simplified for parametric analysis. In the actual gasifier, the parameters such as T_{gas} and m_{dep} would have considerable variations in both the streamwise and angular directions on the wall. These parameters also interact closely with each other, but were assumed to be independent in order to identify their influences on the slag behaviors and underlying mechanisms. The gas temperature (T_{gas}) at the reference condition was fixed at 1800 K along the wall. A change in T_{gas} accompanied the temperature of the ash deposition onto the wall (T_{dep}), which was assumed to be 50 K less. The ash deposition rate (m_{dep}) at the reference condition was 5 kg/s, which was uniformly distributed per area along the wall. T_{dep} was also varied by 10% from the reference value of 1750 K, while T_{gas} remained fixed. As an important parameter for the gasifier shape, the bottom cone height was varied to 0.33 and 0.27 m, while the radius of the slag tap was fixed. The corresponding wall angles were 13.2° and 10.8°, respectively. The height of the main body is another important parameter for the gasifier design, but its effect on the slag behavior was found to be negligible under the reference condition with a fixed m_{dep} for the section.

Table 1. Test parameters varied by ±10% for evaluation of their effects on slag behaviors.

Test Parameters	Reference Value	Accompanying Changes
Gas temp. (T_{gas})	1800 K	T_{dep} (=T_{gas} − 50) changed correspondingly
Ash deposition rate (m_{dep})	5 kg/s	
Ash deposition temp. (T_{dep})	1750 K	
Bottom cone height	0.30 m	With the fixed slag tap radius, the wall angle changed from 12° to 13.2° (+10%) and 10.8° (−10%)
Viscosity of liquid slag (μ)	Equation (6)	T_{cv} changed from 1548 K to 1557 K (+10%) and 1538 K (−10%)
Thermal conductivity of slag (k)	Equation (7)	
Emissivity of liquid slag (ε)	0.83	

With regard to the slag properties, the viscosity, thermal conductivity, and emissivity were also varied from the reference conditions. The property submodels were also described in detail in our previous study [21]. The viscosity of the reference condition was estimated using the correlation of Kalmanovich and Frank [23] as follows:

$$\mu = 3.27 \times 10^{-10} T \exp(27420.6/T) \tag{6}$$

The constant term (3.27×10^{-10}) in the above equation was multiplied by 1.1 and 0.9 to vary the viscosity. This was accompanied by a change in T_{cv} at 25 Pa·s from 1548 K to 1538 K and 1558 K, respectively. The thermal conductivity was determined from the thermal diffusivity ($k/\rho C_p$) which was fixed at 4.5×10^{-7} m²/s [24]. The slag density was calculated using the slag composition and found to be 2507 kg/m³ [25]. The specific heat was also calculated using the correlation from the literature [24] and found to be 1.40 kJ/kg·K for the liquid and solid slag above the glass transition temperature (T_{glass}), and $0.922 + 1.796 \times 10^{-4} T - 0.218/T^2$ kJ/kg·K for the solid slag below T_{glass}. T_{glass} was determined to be 992 K for a Cp of 1.1 kJ/kg·K [24]. Therefore, the thermal conductivity became:

$$\text{Above } T_{glass}\text{: } k = 1.58 \text{ (W/m·K)}$$
$$\text{Below } T_{glass}\text{: } k = 1.040 + 2.025 \times 10^{-4}T - 0.246/T^2 \text{ (W/m·K)} \tag{7}$$

The surface emissivity of the liquid slag was also varied from 0.83 [24] to 0.747 and 0.913, which determined the heat transfer rate (Q_{GL}) by radiation.

In the boundary condition for heat transfer analysis, the coolant (water/steam) was assumed to have a fixed temperature of 523 K under evaporation with a heat transfer coefficient of 10^4 W/m²·K. The membrane tube for the coolant had a thermal conductivity of 43 W/m·K and thickness of 6.3 mm. The refractory lining between the membrane tube and solid slag had a thermal conductivity of 8 W/m·K and thickness of 16.0 mm [22].

To assess the influence of the parameters variation, the results were summarized for: (I) the thicknesses of the liquid slag (δ_L) and solid slag (δ_S) at the slag tap; (II) the total heat transfer rates from the gas to the liquid slag (Q_{GL}) and to the solid slag (Q_{LS}); and (III) the temperatures on the liquid slag surface (T_{surf}) and refractory-solid slag interface (T_R) at the slag tap.

3. Results and Discussion

3.1. Effects of Gas Temperature

Figure 2 shows the contours of the temperature, viscosity, and velocity in the liquid slag layer for the reference value of T_{gas} and its variation of ±10%. Regardless of the value of T_{gas}, δ_L exhibited a similar trend along the wall influenced by the gasifier geometry. The liquid slag rapidly built up at the top cone, and gradually increased in the vertical main body with the continuous deposition of the ash. Once it entered the bottom cone, δ_L suddenly increased because of the change in the wall angle (α) from 90° to 12°. This illustrates a large influence of the wall angle at the bottom cone, which is evaluated in detail later. The value of δ_L at the slag tap (y = 0 m) was 17.4 mm for T_{gas} = 1800 K, 21.6 mm for 1620 K, and 14.4 mm for 1980 K.

Within the temperature contours (Figure 2a–c), the interface temperature with the solid slag remained fixed at T_{cv} (1548 K), but the surface temperature (T_{surf}) facing the gas was influenced by T_{gas}. The gap between the isothermal lines was uniform across the layer, indicating that the temperature profile in the liquid slag had close to a linear relationship. Based on the correlation in Equation (6), the viscosity of the liquid slag exponentially decreased from the interface to the surface facing the syngas as shown in Figure 2d–f. Because of the reduced viscosity near the surface, the liquid slag flowed downward more quickly (Figure 2g–i). For example, the surface velocity of the reference case (Figure 2e) reached 0.045 m/s at the end of the main body of the reactor, but was reduced to as low as 0.028 m/s when the slag flow entered the bottom cone region (y = 0.295 m). The velocity was then restored to 0.075 m/s at y = 0 m as the temperature near the surface further increased as a result of the continuous heat input from the hot syngas.

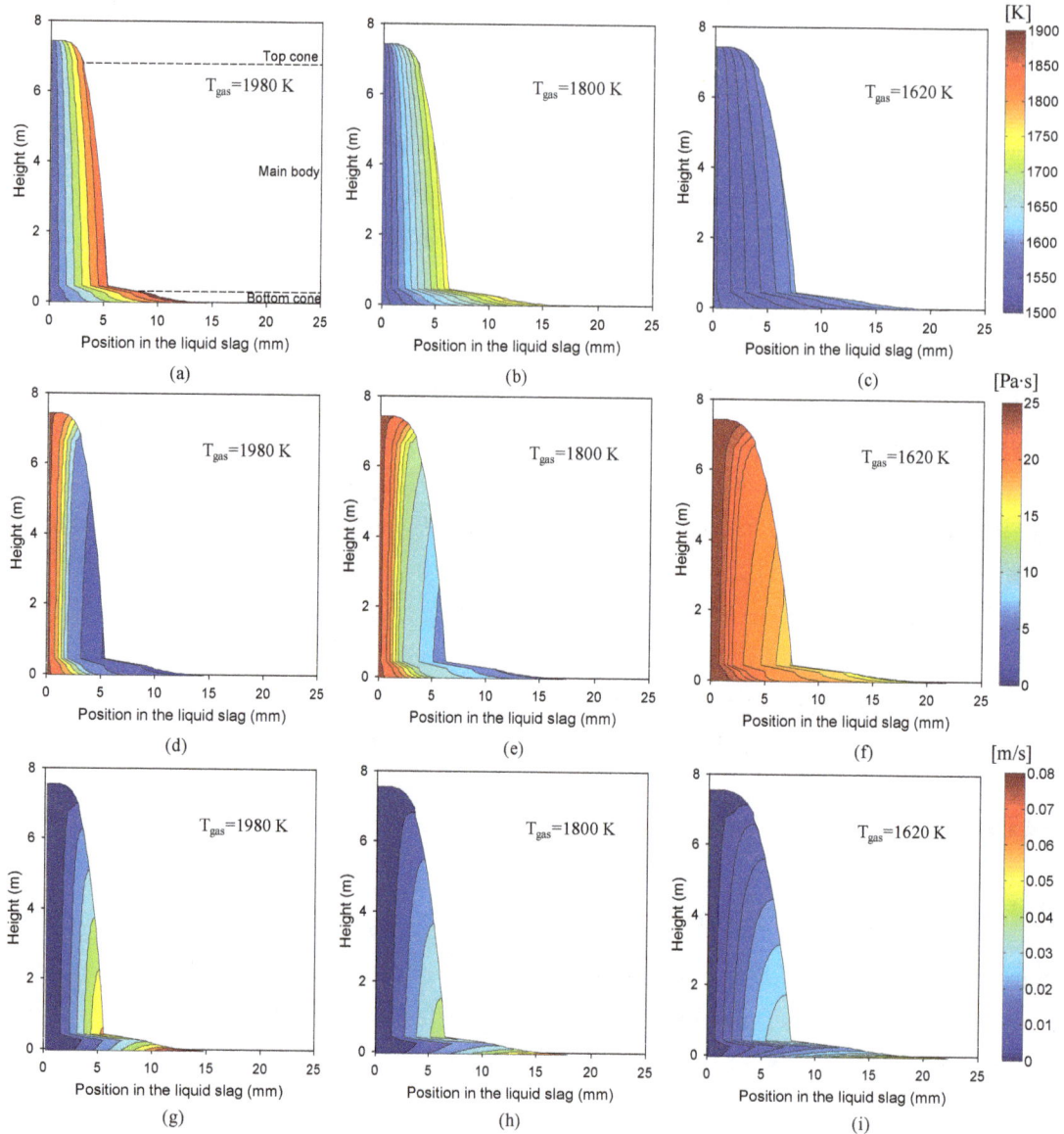

Figure 2. Slag temperature (**a–c**), viscosity (**d–f**) and velocity distribution (**g–i**) in the liquid slag layer for different gas temperature conditions.

Table 2 lists the key results at the slag tap for the ±10% variations in T_{gas} from the reference value (1800 K). Here, Q_{GL} and Q_{LS} represent the total heat transfer rate (kW) from the gas to the liquid and solid slag, respectively, integrated along the wall. The slag thicknesses and interface temperatures (T_{surf} and T_R) were evaluated at the slag tap. T_{gas} was found to have the greatest influence on the slag flow and heat transfer. Both Q_{GL} and Q_{LS} were approximately doubled by a 10% increase in T_{gas} and reduced to a quarter by a 10% decrease, because radiation ($\sim T^4$) was the main mode of heat transfer at such a high temperature. δ_S was inversely proportional to q_{LS} ($=Q_{LS}/A$), as indicated in Equation (5) for the thermal conduction across the solid slag layer. Therefore, it gradually increased as q_{LS} decreased along the wall.

Table 2. Changes in key output values for ±10% variations in input parameters from the reference case (Q_{GL} and Q_{LS}: integrated over the entire wall; δ_L, δ_S, T_{surf}, and T_R: at the slag tap).

Varied Parameters		Q_{GL} (%)	Q_{LS} (%)	δ_L (%)	δ_S (%)	T_{surf} (K)	T_R (K)
Gas temperature (T_{gas})	+10%	106.4	107.1	−17.4	−54.8	168.6	50.1
	−10%	−74.3	−78.2	24.0	405.5	−167.5	−35.7
Ash deposition rate (m_{dep})	+10%	−2.6	−2.0	3.2	3.1	0.5	−1.3
	−10%	2.9	2.3	−3.4	−3.4	−0.5	1.5
Ash deposition temp. (T_{dep})	+10%	−16.7	6.1	−1.2	−5.8	10.5	2.7
	−10%	16.6	−6.1	1.2	6.6	−10.6	−2.7
Bottom cone height	+10%	2.3	3.4	−2.9	−2.7	−0.5	1.1
	−10%	−2.4	−3.6	3.3	3.2	0.6	−1.3
Liquid slag viscosity (μ)	+10%	−4.1	−4.3	1.2	6.0	1.0	−2.0
	−10%	4.6	4.7	−1.3	−6.1	−1.1	2.3
Slag conductivity (k)	+10%	7.1	6.9	0.3	1.3	−2.0	3.7
	−10%	−7.5	−7.3	−0.3	−1.3	2.0	−3.7
Liquid slag emissivity (ε)	+10%	2.8	2.4	−0.3	−1.7	2.0	0.8
	−10%	−3.2	−2.8	0.4	2.1	−2.4	−0.9
Values in reference case		179.4 kW	182.6 kW	17.4 mm	69.3 mm	1777.2 K	567.1 K

As shown in Table 2, δ_L was less influenced by T_{gas} than δ_S. For example, it increased by 24% for a T_{gas} value of 1620 K. This was because the change in δ_L was moderated by the velocity profile. As shown in Figure 2f, the low value of T_{gas} greatly increased the slag viscosity (e.g., from 2.9 Pa·s at the surface at y = 0 m for T_{gas} = 1800 K to 19.2 Pa·s for T_{gas} = 1620 K). However, the corresponding decrease in the surface velocity was only 0.017 m/s. The velocity close to the interface of the solid and liquid slag (r = 0 mm) remained unaffected owing to the no-slip condition and the critical viscosity being fixed at 25 Pa·s. Because such velocity profiles determine the thickness (m ~ $\rho\delta_L v_{avg}$), δ_L was less sensitive to the changes in T_{gas}.

To investigate the exact trend in the slag thicknesses, additional simulations were carried out for two intermediate values of T_{gas} at 1710 K and 1890 K. Figure 3 summarizes the results of δ_L, δ_S, and T_{surf} at the slag tap for different values of T_{gas}. Here, δ_S exponentially increased with a decrease in T_{gas}. Under the extreme condition of T_{gas} = 1620 K, the solid slag became as thick as 350 mm at the slag tap, which would be thick enough to block the slag tap. Because δ_S was larger and more sensitive to the change in T_{gas} than δ_L, this is a crucial parameter of the slag behaviors in relation to the prevention of blockage at the slag tap.

Figure 3 also shows that T_{surf} exhibited a linear relationship with T_{gas}, which can be explained by the overall energy balance. For a section of the liquid slag layer, the energy balance can be approximated as follows:

$$\varepsilon\sigma A\left(T_{gas}^4 - T_{surf}^4\right) \cong kA\frac{T_{surf} - T_{cv}}{\delta_L} + \left(H_{out} - H_{in}\right)_{total} \tag{8}$$

The terms in Equation (8) represent the radiative heat transfer from the gas (Q_{GL}), conduction through the liquid slag layer, and the enthalpy difference between input and output flow, respectively. When rearranged for T_{gas} and T_{surf}:

$$T_{gas}^4 \cong T_{surf}^4 + k \frac{T_{surf} - T_{cv}}{\varepsilon\sigma\delta_L} + \frac{\left(H_{out} - H_{in}\right)_{total}}{\varepsilon\sigma A} \qquad (9)$$

The second and third terms in the RHS of Equation (9) were two-orders of magnitude smaller than those of T_{gas}^4 and T_{surf}^4. Therefore, T_{surf} changed almost linearly with the variations of T_{gas}. However, the difference between the two temperatures ($T_{gas} - T_{surf}$) increased from 10 K for T_{gas} = 1620 K to 34 K for T_{gas} = 1980 K.

With regard to the influence of low gas temperatures on the slag flow, it is worth reiterating the results from our previous study [21]. The gas temperature at the bottom cone of this gasifier is typically lower than at main body part where partially oxidative reactions of coal take place. If T_{gas} suddenly falls from 1800 K to below T_{cv} at the bottom cone, the hot liquid slag from the main body of the gasifier acts as a temporary heat reservoir, providing heat to the solid slag and gas at both ends. This prevents immediate increases in both δ_L and δ_S in the bottom cone.

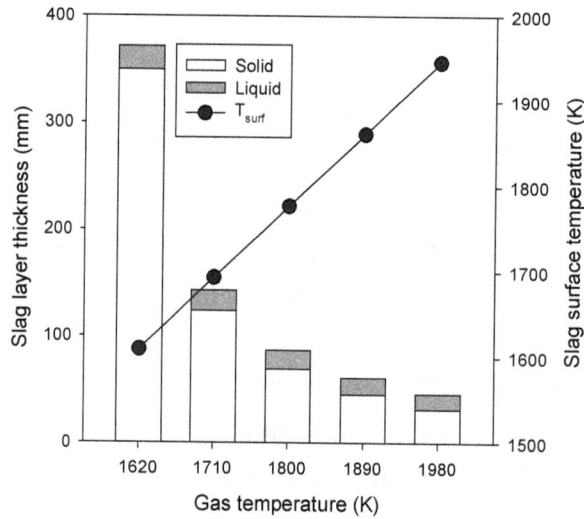

Figure 3. Effects of gas temperature on the slag thicknesses and surface temperature at the slag tap.

3.2. Effects of Ash Deposition Rate

When the particle deposition rate (m_{dep}) changes, it would be reasonable to expect that the slag thickness, especially δ_L, would be affected by a similar magnitude. However, the values listed in Table 2 show that the changes in the slag thicknesses were reduced to approximately one third of the variation in m_{dep}. This can be explained by the velocity distribution in the liquid slag layer plotted in Figure 4. When m_{dep} was increased by 10%, the liquid slag near the surface was accelerated by 0.005 m/s, and δ_L was extended by 0.5 mm. Considering that the area under the curve represents the mass flow rate, the increased portion in m_{dep} or the area was absorbed near the surface, where the liquid slag flowed faster. Therefore, δ_L became less sensitive to the variations in m_{dep}. Q_{GL} and Q_{LS} corresponded to the change in δ_L, and were within 3% of their values in the reference case (Table 2).

Figure 4 also shows that the temperature profiles in the liquid slag at the slag tap remained linear, and the changes in T_{surf} were very small (0.5 K). Although the temperature gradient reflected the change in δ_L, the gradient at the interface to the solid slag (*i.e.*, Q_{LS}) was slightly less sensitive. This led to a 0.6% smaller change in Q_{LS}, compared to that in Q_{GL}.

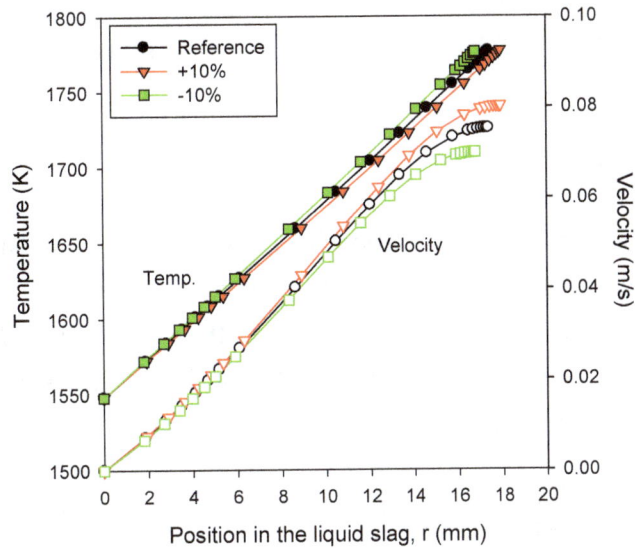

Figure 4. Effects of ash deposition rate on temperature and velocity profiles of liquid slag at the slag tap.

3.3. Effects of Ash Deposition Temperature

In an entrained flow gasifier, inert particles almost immediately reach thermal equilibrium with the gas. However, the particle deposition temperature (T_{dep}) can be influenced by the exothermic oxidation of char or endothermic gasification reactions. When T_{dep} was varied to 1575 or 1825 K from 1750 K, its impact was noticeable in T_{surf} as shown in Figure 5.

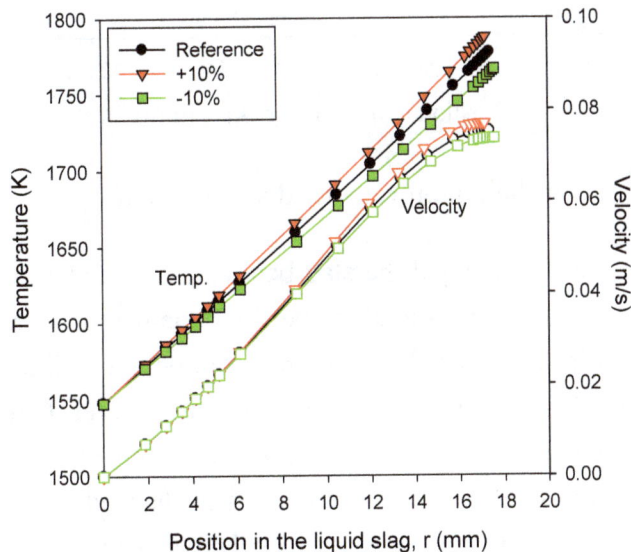

Figure 5. Effects of ash deposition temperature on temperature and velocity profiles of liquid slag at the slag tap.

For example, T_{surf} became 10.5 K higher by the deposition of hotter ash. Because this decreased the temperature difference with the gas ($T_{gas}^4 - T_{surf}^4$), Q_{GL} was decreased by 16.7% compared to the reference case. In contrast, Q_{LS} was increased by 6.1% by the higher T_{surf}. Among the parameters investigated in this study, T_{dep} was a unique parameter that caused opposite trends in Q_{GL} and Q_{LS}. Despite the temperature change near the surface, the velocity profile shown in Figure 5 was not sensitively changed because the viscosity was already low at such high temperatures. This led to a very small change (1.2%) in δ_L.

3.4. Effects of Bottom Cone Design

The bottom cone height of the gasifier was varied by 10% from 0.3 m for a fixed slag tap radius. This corresponded to a change in the wall angle of ±1.2°. The values listed in Table 2 show that the influence was approximately 3%, which was smaller than in the other cases. When the bottom cone height was decreased to 0.27 m, for example, the liquid slag at the surface was slowed down by only 0.0025 m/s (Figure 6). This led to an increase in δ_L of 2.9%.

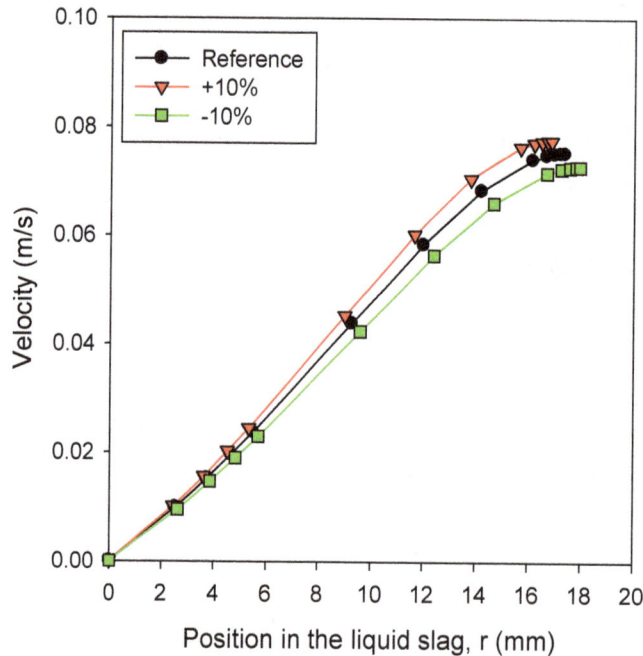

Figure 6. Effects of bottom cone height on velocity distribution of liquid slag at the slag tap.

Since the bottom cone immediately changed the slag behaviors as shown in Figure 2, the effect of the bottom cone design required further investigations before conclusions were reached. The Prenflo gasifier considered in this study already has a short bottom cone, with a wall angle of 12°. In contrast, recent designs for Shell gasifiers reported in the literature have larger bottom cone angles [26]. Therefore, additional cases were studied, in which the wall angle was changed to 18°, 24°, and 30°. The slag tap radius and ash deposition in the bottom cone remained unchanged. Figure 7 shows the velocity distribution within the liquid slag at the slag tap for these cases. The velocity was increased from 0.075 m/s for 12° to 0.102 m/s for 30° by the increased gravity force in the streamwise direction. This reduced δ_L to 13.2 mm. Table 2 compares the slag thicknesses for the additional cases. Both δ_L and δ_S were reduced by 24% for 30°. More importantly, the equivalent thickness in the horizontal direction

$((\delta_L + \delta_S)/\cos \alpha)$ was reduced to about a third. The results clearly suggest that the bottom cone angle is a crucial parameter in reducing the possibility of blockage at the slag tap.

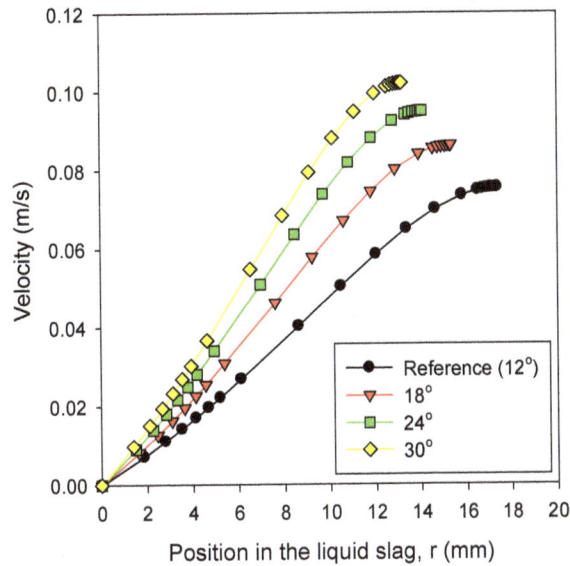

Figure 7. Effects of bottom cone angle on velocity distribution of liquid slag at the slag tap.

3.5. Effects of Slag Properties: Viscosity

The viscosity of the liquid slag is a crucial property governing the slag behaviors on the wall. Because of the difficulty in measuring the viscosity at high temperatures, several correlations based on the slag composition have been proposed [23,27–30]. However, these correlations exhibited deviations from the measured data depending on the ash samples [31–33] and with each other [34]. Predicting the influence of the viscosity on the slag behaviors would be helpful in evaluating the impact of the uncertainties involved in the correlations. This is also meaningful in terms of the gasifier operation, because the ash characteristics are important parameters in determining the range of suitable coals and the amount of flux required [11].

The $\pm 10\%$ variations in the slag viscosity were accompanied by changes in T_{cv} from 1548 K to 1557 K ($+10\%$) and 1538 K (-10%), because the critical viscosity was assumed to be constant at 25 Pa·s. The values listed in Table 2 show that the increased viscosity thickened the liquid slag by only 1.2%, and the change in T_{surf} was very small (1 K). Because the viscosity at $r = 0$ m remained the same, the influence on the velocity was visible toward the surface, with changes of about 0.0015 m/s as shown in Figure 8a. With a change in T_{cv} of about 10 K, however, the temperature gradient within the slag layer (i.e., the heat transfer rate) was reduced as shown in Figure 8b. This resulted in a 4%–5% change in both Q_{GL} and Q_{LS} (Table 2). As determined by Fourier's law ($\delta_S \sim (T_{cv} - T_R)/Q_{LS}$), δ_S was influenced not only by Q_{LS} but also by T_{cv}. This led to a change in δ_S of approximately 6%, which was about five times larger than that in δ_L.

However, the above results require careful interpretation because the numerical model assumed the steady-state condition. A transient change in the viscosity of fresh liquid slag does not immediately affect T_{cv} at the inner layer facing the solid slag. If T_{cv} and T_{gas} remain the same, the results indicate that the change in the slag thickness would be very small. Therefore, expanding the model for transient simulations would be helpful in understanding the time-scale for the impact of the changes in the slag viscosity on the slag thickness. In the long term, lowering the slag viscosity (by changing the ash composition or the injection of flux) would reduce δ_S more than δ_L by the change in T_{cv}, as indicated in the results.

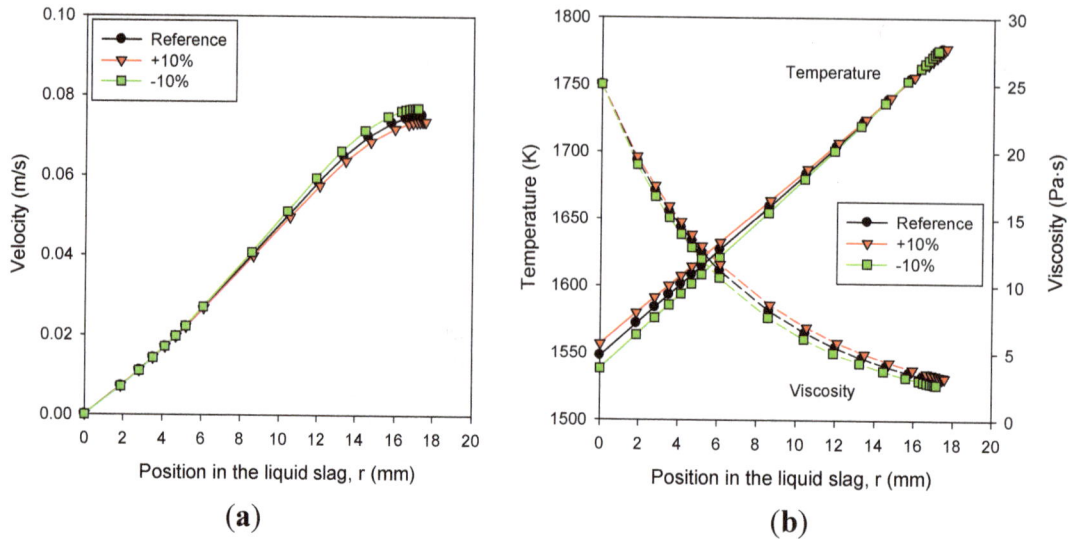

Figure 8. Effects of slag viscosity on **(a)** velocity; **(b)** temperature and viscosity distribution of liquid slag at the slag tap.

3.6. Effects of Slag Properties: Thermal Conductivity and Emissivity

The thermal conductivity of the slag positively influenced the heat transfer rates by Fourier's law. The results in Table 2 show that Q_{GL} and Q_{LS} were changed by approximately ±7% for 10% variations in the thermal conductivity. However, the values of T_{surf} and δ_L remained almost unaffected, with changes of approximately 2 K and 0.3%, respectively. Such a trend for T_{surf} was observed along the entire wall, as shown in Figure 9.

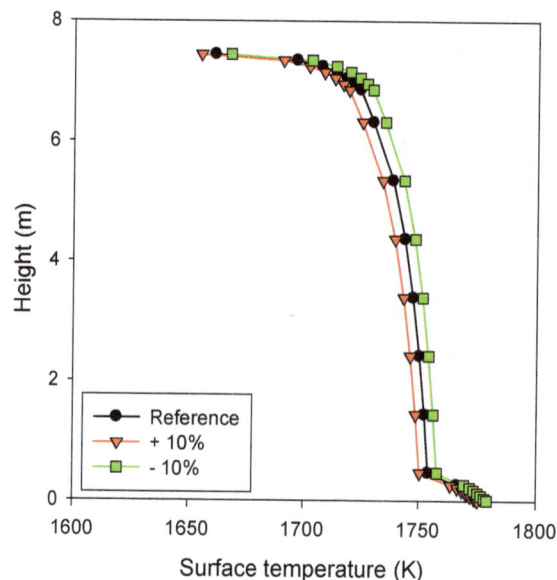

Figure 9. Effects of slag conductivity on surface temperature of liquid slag along the wall.

Compared to δ_L, the changes in δ_S were larger (1.3%). This was because a relatively larger change in T_R (3.7 K) was also induced by Q_{LS}. Overall, the effect of the thermal conductivity on Q_{LS} (to the coolant in the wall) was larger than that of the other parameters, except for T_{gas}. In contrast, its effect on the slag behaviors was smaller.

The surface emissivity (ε) of the liquid slag was found to have less influence on the slag behaviors, as presented in Table 2. Although Q_{GL} is proportional to ε based on the Stefan–Boltzmann equation, the actual changes in Q_{GL} were only about 3% for $\pm10\%$ variations in ε from 0.83. This was because it accompanied a change in T_{surf}. When ε was decreased to 0.747, for example, T_{surf} was also lowered by 2.4 K (Table 2). Since this contributed toward an increase in Q_{GL}, the resultant decrease in Q_{GL} was limited to 3.2%. Subsequently, δ_S was increased by 2.1%. The change in δ_L was very small because the temperature and viscosity within the liquid slag layer were only slightly influenced.

4. Conclusions

Using the numerical model for slag flow, the influences of the design/operation parameters and slag properties were investigated for a commercial coal gasifier by varying the parameters by $\pm10\%$ from the reference conditions. The key findings are as follows.

- The velocity profile of the liquid slag was less sensitive to the variations in the parameters, and therefore, the change in the thickness of the liquid slag was typically smaller than that of the solid slag.
- The gas temperature was found to be highly influential, because of its dominant effect on the radiative heat transfer to the slag layer. The solid slag thickness increased exponentially with an increase in the gas temperature.
- The effect of the variations in the ash deposition rate was diminished by the high-velocity region developed near the liquid slag surface. Increasing the ash deposition rate by 10% caused an approximate 3% increase in the thickness of the slag layers.
- The slag viscosity significantly influenced the solid slag thickness through the corresponding changes in the temperature (T_{cv}) and its gradient (heat flux) at the interface of the solid and liquid slag layers. Decreasing the slag viscosity by 10% reduced the thickness of the liquid slag by only 1.3%, whereas that of the solid slag was reduced by 6%.
- A higher thermal conductivity of the slag directly increased the heat transfer rate across the slag layer, whereas its effect on the thickness of the slag layers was very small.
- For the bottom cone of the gasifier, steeper angles were favorable to reduce the slag layer thickness.

In an actual gasifier, the reactions and heat transfer in the gasifier and the slag behaviors on the wall are closely coupled and interact with each other, unlike the simplification in this parametric study. Therefore, applications of the numerical model integrated with a process simulation or computational fluid dynamics are required to gain a deeper understanding of the complex interactions.

Acknowledgments

This work was supported by the New & Renewable Energy Core Technology Program of the Korea Institute of Energy Technology Evaluation and Planning (KETEP) and Doosan Heavy Industries and Construction granted financial resource from the Ministry of Trade, Industry & Energy, Korea (2011951010001A).

Author Contributions

Changkook Ryu and Insoo Ye formulated the numerical model for the slag, and Insoo Ye and Junho Oh developed the Excel VBA code for the model. All authors were involved in determining the simulation conditions, analyzing the results and preparing the manuscript.

Nomenclature

A	area, m^2		g	gravity, 9.81 m/s^2
H	enthalpy, J/s		k	thermal conductivity, $W/m \cdot K$
M	momentum, $kg \cdot m/s^2$		m	mass flow rate, kg/s
Q	heat transfer rate, W		q	heat flux, W/m^2
r	radius perpendicular to the wall, m		T	temperature, K
V	volume, m^3		v	streamwise velocity, m/s
y	length parallel to the wall, m			

Greek

α	angle from the horizontal plane °		δ	thickness of a slag layer, m
ε	emissivity		μ	viscosity, $Pa \cdot s$
ρ	density, kg/m^3			

Subscript

cond	conduction		cv	critical viscosity
dep	depositing slag		gas	gas
GL	from gas to liquid slag		glass	glass transition of slag
in	inflow		L	liquid slag
LS	from liquid slag to solid slag		out	outflow to the section below
react	reactions of residual carbon or the phase transformation		R	refractory
S	solid slag		surf	liquid slag surface facing gas

Conflicts of Interest

The authors declare no conflicts of interest.

References

1. Stiegel, G.J.; Maxwell, R.C. Gasification technologies: The path to clean, affordable energy in the 21st century. *Fuel Process. Technol.* **2001**, *71*, 79–97.

2. Higman, C.; van der Burgt, M. *Gasification*, 2nd ed.; Elsevier: New York, NY, USA, 2008.

3. Wang, P.; Massoudi, M. Slag behavior in gasifiers. Part I: Influence of coal properties and gasification conditions. *Energies* **2013**, *6*, 784–806.

4. Duchesne, M.A.; Hughes, R.W.; Lu, D.Y.; McCalden, D.J.; Anthony, E.J.; Macchi, A. Fate of inorganic matter in entrained-flow slagging gasifiers: Pilot plant testing. *Fuel Process. Technol.* **2014**, *125*, 18–33.

5. Ni, J.J.; Zhou, Z.J.; Yu, G.S.; Liang, Q.F.; Wang, F.C. Molten slag flow and phase transformation behaviors in a slagging entrained-flow coal gasifier. *Ind. Eng. Chem. Res.* **2010**, *49*, 12302–12310.

6. Ilyushechkin, A.Y.; Hla, S.S.; Roberts, D.G.; Kinaev, N.N. The effect of solids and phase compositions on viscosity behaviour and Tcv of slags from Australian bituminous coals. *J. Non-Cryst. Solids* **2011**, *357*, 893–902.

7. Song, W.J.; Dong, Y.H.; Wu, Y.Q.; Zhu, Z.B. Prediction of temperature of critical viscosity for coal ash slag. *AIChE J.* **2011**, *57*, 2921–2925.

8. McLennan, A.R.; Bryant, G.W.; Bailey, C.W.; Stanmore, B.R.; Wall, T.F. An experimental comparison of the ash formed from coals containing pyrite and siderite mineral in oxidizing and reducing conditions. *Energy Fuels* **2000**, *14*, 308–315.

9. McLennan, A.R.; Bryant, G.W.; Bailey, C.W.; Stanmore, B.R.; Wall, T.F. Index for iron-based slagging for pulverized coal firing in oxidizing and reducing conditions. *Energy Fuels* **2000**, *14*, 349–354.

10. Duchesne, M.A.; Hall, A.D.; Hughes, R.W.; McCalden, D.J.; Anthony, E.J.; Macchi, A. Fate of inorganic matter in entrained-flow slagging gasifiers: Fuel characterization. *Fuel Process. Technol.* **2014**, *118*, 208–217.

11. Yun, Y.; Yoo, Y.D.; Chung, S.W. Selection of IGCC candidate coals by pilot-scale gasifier operation. *Fuel Process. Technol.* **2007**, *88*, 107–116.

12. Song, W.; Tang, L.; Zhu, X.; Wu, Y.; Rong, Y.; Zhu, Z. Fusibility and flow properties of coal ash and slag. *Fuel* **2009**, *88*, 297–304.

13. Kong, L.; Bai, J.; Bai, Z.; Guo, Z.; Li, W. Improvement of ash flow properties of low-rank coal for entrained flow gasifier. *Fuel* **2014**, *120*, 122–129.

14. Kong, L.; Bai, J.; Li, W.; Bai, Z.; Guo, Z. Effect of lime addition on slag fluidity of coal ash. *J. Fuel Chem. Technol.* **2011**, *39*, 407–411.

15. Kong, L.; Bai, J.; Bai, Z.; Guo, Z.; Li, W. Effects of $CaCO_3$ on slag flow properties at high temperatures. *Fuel* **2013**, *109*, 76–85.

16. Wang, J.; Liu, H.; Liang, Q.; Xu, J. Experimental and numerical study on slag deposition and growth at the slag tap hole region of Shell gasifier. *Fuel Process. Technol.* **2013**, *106*, 704–711.

17. Roberts, D.G.; Harris, D.J.; Tremel, A.; Ilyushechkin, A.Y. Linking laboratory data with pilot scale entrained flow coal gasification performance. Part 2: Pilot scale testing. *Fuel Process. Technol.* **2012**, *94*, 26–33.

18. Sun, B.; Liu, Y.; Chen, X.; Zhou, Q.; Su, M. Dynamic modeling and simulation of shell gasifier in IGCC. *Fuel Process. Technol.* **2011**, *92*, 1418–1425.

19. Yang, Z.; Wang, Z.; Wu, Y.; Wang, J.; Lu, J.; Li, Z.; Ni, W. Dynamic model for an oxygen staged slagging entrained flow gasifier. *Energy Fuels* **2011**, *25*, 3646–3656.

20. Yong, S.Z.; Gazzino, M.; Ghoniem, A. Modeling the slag layer in solid fuel gasification and combustion—Formulation and sensitivity analysis. *Fuel* **2012**, *92*, 162–170.

21. Ye, I.; Ryu, C. Numerical modeling of slag flow and heat transfer on the wall of an entrained coal gasifier. *Fuel* **2015**, *150*, 64–74.

22. Seggiani, M. Modeling and simulation of time varying slag flow in a Prenflo entrained-flow gasifier. *Fuel* **1998**, *77*, 1611–1621.

23. Kalmanovitch, D.P.; Frank, M. An effective model of viscosity for ash deposition phenomena. In Proceedings of the Mineral Matter and Ash Deposition from Coal, Santa Barbara, CA, USA, 22–26 February 1998; Bryers, R.W., Vorres, K.S., Eds.; United Engineering Trustees Inc.: New York, NY, USA, 1998.

24. Mills, K.C.; Rhine, J.M. The measurement and estimation of the physical properties of slag formed during coal gasification: 2. Properties relevant to heat transfer. *Fuel* **1989**, *68*, 904–910.

25. Mills, K.C.; Rhine, J.M. The measurement and estimation of the physical properties of slag formed during coal gasification: 1. Properties relevant to fluid flow. *Fuel* **1989**, *68*, 193–200.

26. Liu, S.; Tao, M.; Hao, Y. A numerical model for chemical reaction on slag layer surface and slag layer behavior in entrained-flow gasifier. *Therm. Sci.* **2013**, *17*, 1389–1394.

27. Watt, J.D.; Fereday, F. The flow properties of slags formed from the ashes of British coals: Part 1. viscosity of homogeneous liquid slags in relation to slag composition. *J. Inst. Fuel* **1969**, *42*, 99–103.

28. Reid, W.T. *External Corrosion and Deposits: Boilers and Gas Turbines*; American Elsevier Publishing Co.: New York, NY, USA, 1971.

29. Browning, G.J.; Bryant, G.W.; Hurst, H.J.; Lucas, J.A.; Wall, T.F. An empirical method for the prediction of coal ash slag viscosity. *Energy Fuels* **2003**, *17*, 731–737.

30. Urbain, G. Viscosity estimation of slags. *Steel Res. Int.* **1987**, *58*, 111–116.

31. Zhang, L.; Jahansahi, S. Review and modeling of viscosity of silicate melts: Part I. Viscosity of binary and ternary silicates containing CaO, MgO, and MnO. *Metall. Mater. Trans. B* **1998**, *29*, 177–186.

32. Kondratiev, A.; Jak, E. Review of Experimental data and modeling of viscosities of fully liquid slags in the Al2O3-CaO-'FeO'-SiO2 system. *Metall. Mater. Trans. B* **2001**, *32*, 1015–1025.

33. Seetharaman, S.; Mukai, K.; Sichen, D. Viscosities of slags—An overview. *Steel Res. Int.* **2005**, *76*, 267–278.

34. Duchesne, M.A.; Bronsch, A.M.; Hughes, R.W.; Masset, P.J. Slag viscosity modeling toolbox. *Fuel* **2013**, *114*, 38–43.

5

Systematic Methods for Working Fluid Selection and the Design, Integration and Control of Organic Rankine Cycles—A Review

Patrick Linke [1,*], Athanasios I. Papadopoulos [2,†] and Panos Seferlis [3,†]

[1] Department of Chemical Engineering, Texas A&M University at Qatar, P.O. Box 23874, Education City, 77874 Doha, Qatar

[2] Chemical Process and Energy Resources Institute, Centre for Research and Technology-Hellas, Thermi, 57001 Thessaloniki, Greece; E-Mail: spapadopoulos@cperi.certh.gr

[3] Department of Mechanical Engineering, Aristotle University of Thessaloniki, P.O. Box 484, 54124 Thessaloniki, Greece; E-Mail: seferlis@auth.gr

† These authors contributed equally to this work.

* Author to whom correspondence should be addressed; E-Mail: patrick.linke@qatar.tamu.edu

Academic Editor: Roberto Capata

Abstract: Efficient power generation from low to medium grade heat is an important challenge to be addressed to ensure a sustainable energy future. Organic Rankine Cycles (ORCs) constitute an important enabling technology and their research and development has emerged as a very active research field over the past decade. Particular focus areas include working fluid selection and cycle design to achieve efficient heat to power conversions for diverse hot fluid streams associated with geothermal, solar or waste heat sources. Recently, a number of approaches have been developed that address the systematic selection of efficient working fluids as well as the design, integration and control of ORCs. This paper presents a review of emerging approaches with a particular emphasis on computer-aided design methods.

Keywords: organic Rankine cycle; systematic approaches; design; optimisation; working fluid selection

1. Introduction

Over the past decade increasing concerns over climate change and high energy prices have resulted in a strong interest to utilize waste or renewable heat sources for power generation. For such applications the Organic Rankine Cycle (ORC) is a widely used technology with many installations converting a number of heat sources to power in the kW and MW range [1]. The success of the system is largely attributable to its simplicity and flexibility: ORCs are simple Rankine cycles similar to those used in conventional power plants. ORCs are flexible and can be applied on a wide range of heat source temperatures ranging from 80 to 400 °C [2]. They enable cost efficient power generation from a broad range of heat sources by replacing water with organic working fluids such as refrigerants and other organic molecules to achieve better efficiencies [3]. Although ORCs constitute a proven technology with more than 1.5 GW$_{el}$ of capacity installed world-wide in a variety of applications, including industrial waste heat recovery [1], geothermal [4–6], solar thermal [7] and biomass power plants [8], the key research challenges remain: the identification of high-performance working fluids, the corresponding optimal design configuration and operating characteristics of the thermodynamic cycle and the optimum integration of the ORCs with the available heat sources.

The design challenge is due to the very large number of working fluid chemistries as well as structural and operating ORC parameters that need to be considered as decision options within a systematic problem formulation to identify highly performing systems. The examination of various working fluids or alternative ORC configurations can lead to useful insights regarding potential performance improvements, yet most published works lack the use of systematic methods, rather relying on empirically identified enhancements approached through heuristic treatment of decision options. Empirical investigations are often based on knowledge gained from either experimental or theoretical work and are clearly useful. Yet the proposed improvements may be limited unless an extensive number of ORC working fluid and system characteristics are systematically taken into account during optimization.

Computer-aided technology is a promising tool to exploit empirical know-how and guide the search for novel and efficient technologies as it is capable to cope with the investigation of an enormous range of options [9]. Process design and optimization [10], process integration [11], control [12], molecular design [13–15] and integrated process and molecular design [16] are examples of systematic computer-aided methods with applications in diverse process systems.

Research and development efforts in ORC have made use of computer-aided tools and methods in the past, but they have rarely been used in a systematic context. The use of systematic and robust computer-aided methods in the development and operation of ORC technologies has emerged in the last few years, but it is still limited compared to the very widespread application of similar tools in other technological sectors. Considering the benefits reaped from the application of such tools in other industries, there is great scope for widening their utilization in ORC. There is currently a very wide community of engineers and scientists in ORC research and development, who are beginning to grasp the benefits resulting from use of systematic computer-aided tools.

This paper aims at presenting a review of available systematic methods for working fluid selection and the design, integration and control of ORCs. The scope of the paper is focused on subcritical ORCs. It provides a structured and organized account of the merits of selected works for more efficient technological developments and identifies areas for further research into computer-aided tools and methods for ORC systems engineering.

2. Overview

The basic ORC process for converting heat from a source stream to power consists of a pump, a turbine, a heat source recovery section and a condenser and uses an organic compound as the working fluid. In the heat recovery section, heat is transferred from the hot source stream to generate high pressure working fluid vapor from which power is generated in a turbine. In the condenser, heat is ejected to a cold utility from the low pressure working fluid vapor obtained from the turbine outlet. The resulting liquid working fluid is repressurized in the pump and the cycle is closed in the heat recovery section.

The overall design goal is to maximize ORC system performance for a given situation in terms of heat sources and heat sinks. Different performance criteria have been used as design objectives in ORC studies [17], the most common being the thermodynamic metrics thermal efficiency and exergetic efficiency, with economic criteria such as the power production cost being less common due to the difficulties of precise cost estimation.

The development of a high performance ORC system with respect to any typical performance criterion requires good design choices to be made across the cycle and its interfaces with the heat sources and heat sinks. Important design decisions need to be made with respect to working fluid selection, the cycle design and its operating conditions as well as the heat recovery strategy from the available heat sources, which may involve one or more source streams (Figure 1).

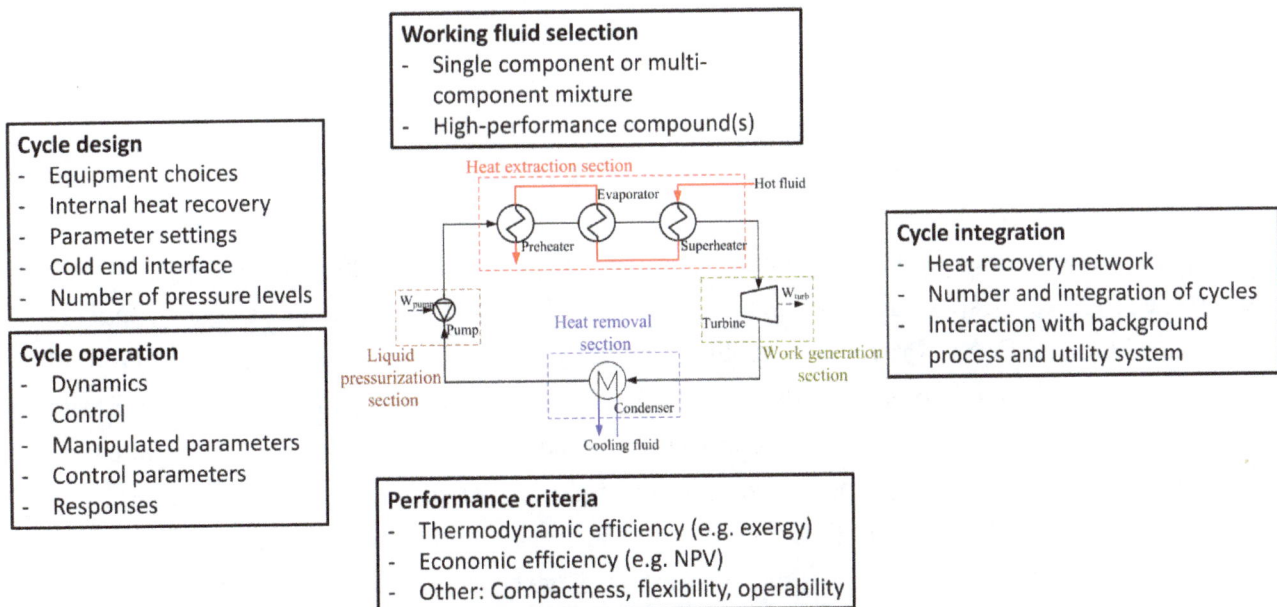

Figure 1. ORC design decisions and objectives.

Many design alternatives exist at each level. For instance, a very large number of alternative working fluids exist, some of which may have never been proposed for ORC systems before. The identification of the best performing working fluid for a given situation requires the ability to systematically screen through the alternatives. Similarly, heat may be transferred from multiple available heat sources so that many alternative strategies may exist to recover the heat into the working fluid. In addition, cycle operating parameters must be optimized to ensure alternatives are compared on the basis of the best possible performance.

Over the years, a number of systematic approaches have emerged to support the designer in optimizing and analyzing alternatives so as to effectively identify the high performing ORC designs for a given problem. These will be reviewed in the remainder of this paper. Section 3 will review systematic approaches to working fluid selection. Section 4 will provide an overview of cycle optimization approaches before approaches to ORC control are reviewed in Section 5. Emerging approaches to the integration of ORCs with multiple heat sources will be reviewed in Section 6.

3. Design and Selection of ORC Working Fluids

3.1. Working Fluid Selection Applications

The choice of ORC working fluids is known to have a significant impact on the thermodynamic as well as economic performance of the cycle. A suitable ORC fluid must exhibit favorable physical, chemical, environmental, safety and economic properties such as low specific volume, viscosity, toxicity, flammability, ozone depletion potential (ODP), global warming potential (GWP) and cost, as well as favorable process attributes such as high thermal and exergetic efficiency, to name but a few [4,5,17–19]. These requirements apply both to pure and mixed working fluids. Existing research is largely focused on the selection of pure working fluids, with well over 100 published reports currently available (see surveys in [1,20]). An important limitation of pure working fluids is their constant temperature profile during phase change [18]. The pinch point encountered at the evaporator and the condenser gives rise to large temperature differences at one end of the heat exchanger leading to high irreversibility. The pinch point is a point of minimum temperature difference between the heat source and the working fluid side of the heat exchanger where the heat transfer is blocked. Working fluid mixtures are more appealing than pure fluids because their evaporation temperature profile is variable, following the profile of the heat source, as opposed to the flat evaporation profile of pure fluids. This enables an approximately stable temperature difference during evaporation, coined as temperature glide, which significantly reduces exergetic losses. Despite their usefulness, the published works addressing the selection of mixed fluids are considerably fewer. Previously published work [21] has investigated different types of multi-component mixtures comprising hydrocarbons, hydrofluorocarbons or siloxanes together with important mixture performance measures and constraints that need to be considered for their evaluation. Hydrocarbon mixtures were also proposed considering regenerative preheating ORC schemes [22], equipment sizing [23] and zeotropic fluids [24] for efficient exploitation of moderate temperature geothermal resources. Mixtures of siloxanes or hydrocarbons have been considered [25] to recover wasted heat from molten carbonate fuel cells using an ORC. Halocarbon mixtures have been investigated [26] for power generation using geothermal heat, indicating significant ORC performance gains compared to pure fluids. A binary mixture of fluorocarbons has been investigated [27] at different concentrations employed in an ORC system for power generation from solar energy. In a similar context, the utilization of hydrocarbon and fluorocarbon mixtures has been investigated [28] at different temperature heat sources in ORCs, evaluating the resulting performance gains using ORC operating parameters like inlet/outlet volume ratio, mass flow, enthalpy difference of expansion *etc*. Different combinations of binary and tertiary mixtures have also been evaluated [29] including alkanes, fluorinated alkanes and siloxanes aiming to find their optimum concentration. An investigation of organic, ammonia-water and alcohol-water mixtures was performed using an optimization method to identify their optimum concentration in ORC and Kalina cycle systems [30]. Mixtures of ammonia- water and CO_2- water were also considered in two new ORC configurations, namely the ORC with liquid-flooded expansion and the ORC with solution circuit [31], with the mixtures employed in the second configuration only. An ammonia-water mixture was also considered in the context of a Kalina cycle and its performance was compared with an ORC using pure ammonia or R134 [32]. Nineteen binary working fluid mixtures were also considered as an alternative to ammonia-water, resulting in the conclusion that the highest performers were propane and propylene-based mixtures. Binary and tertiary polysiloxane mixtures are considered in a different work [33] for ORCs recovering heat from cogeneration plants fed with wood residuals. A zeotropic mixture of R227ea/R245fa is analyzed in a subcritical ORC employed for exploitation of geothermal resources [6].

Whether pure or mixed working fluids, conventional engineering practice mostly considers their selection by testing and comparing various known options from a pre-postulated dataset of several available candidates. As a result, the search is limited to an often arbitrarily compiled list of candidates containing conventional molecules (e.g., refrigerants, hydrocarbons *etc.*). Such a small set is extremely limiting in view of the vast number of molecules that could be considered as candidate ORC working fluids, hence significantly reducing the opportunities for identification of novel and improved options. The limited screening of potential working fluid candidates hampers innovation and a systematic approach is required to enable wider and more systematic searches. The latter is very relevant to the necessity for development of novel chemical compounds which may exhibit favorable characteristics as ORC working fluids and may also overcome the performance of existing ones. In conventional practice this is only possible through experimental work which is clearly useful and irreplaceable. However, experiments involve high costs which are often not justified by the limited performance gains. Computer-aided tools may assist experimental work through their predictive capabilities by systematically guiding searches to options worth investigating. The rather ad-hoc use of such tools in conventional practice prohibits such opportunities.

3.2. Computer-aided Tools: Main Concepts and Challenges

The use of computer-aided tools is clearly very appealing for either the design of novel working fluids or the selection of commercially available ones, in both cases with optimum performance characteristics. The term "design" refers to the determination of a molecular structure regardless of whether such a molecule pre-exists or not. The achievement of optimum performance is rather challenging because it involves two major requirements:

a) The exhaustive generation and evaluation of a very wide range of molecular structures prior to the selection of the working fluid which exhibits a truly optimum performance.

b) The utilization of predictive models which are sufficiently accurate to ensure that the performance of the selected working fluid is both optimum and rigorously validated prior to its practical utilization in an ORC plant.

These requirements are conceptually illustrated in Figure 2 with the aim to provide a comparative assessment of the general predictive model types available to simulate molecular chemistry characteristics and the range of molecules that may be evaluated with each model type within a reasonable computational efficiency. The three model types involve computational chemistry methods, equations of state (EoS) and group contribution (GC) methods in a representation that implies a complementarity due to the existence of shared features that reflect common phenomena and functionalities captured by adjacent models of different abstraction (*i.e.*, rigor of the modelling detail). The simultaneous utilization of representatives from all model types would be ideal as it would satisfy both previous requirements for the identification of optimum and immediately applicable working fluids, yet it would have a detrimental effect on computational efficiency. This is also the main reason why each type of model may only be used independently but not all types are suitable for the optimum design and selection of highly performing working fluids.

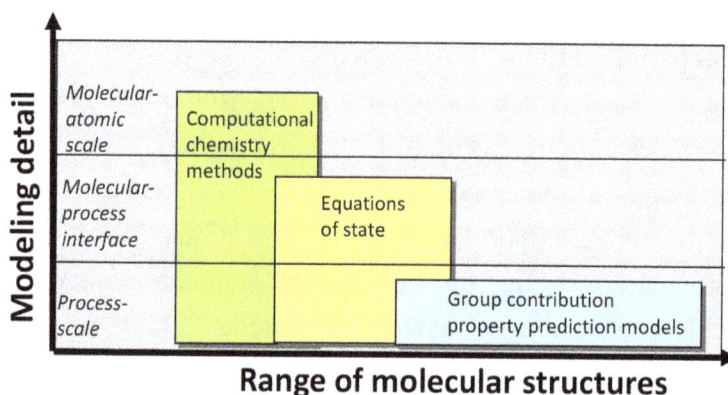

Figure 2. Property prediction models with respect to modeling detail and range of molecular structures which may be simulated at a reasonable computational effort.

Computational chemistry methods [34] involve several different techniques such as density functional theory (DFT) and the conductor-like screening model for real solvents (COSMO-RS) [35] which is based on quantum chemistry and addresses liquid phase predictions. Such methods are based on a robust representation of the molecular chemistry hence they enable the determination of property features at even the atomic or molecular scale. However, the resulting predictions may not be easily transferred into molecular parameters which are required to performed mass and energy balances or to determine operating conditions at the ORC process level. Furthermore, the simulation of even one molecule often requires very extensive computational effort which may range from a few hours to a few days hence prohibiting their use to evaluate the vast number of molecules that may be considered as ORC working fluids. On the other hand, EoS act as an interface between molecular characteristics and process-level properties, while the required computational effort is sufficiently low to use them in an extensive evaluation of working fluids. However, chemical or physical parameters required as inputs to characterize molecular or mixture behaviors are available for relatively few molecules, prohibiting the direct and wide utilization of EoS in the design and selection of working fluids.

GC methods [36] avoid the bottlenecks of computational efficiency and data unavailability because they are based on relatively simpler (hence computationally faster) predictive models than the other two methods, while they refer to molecular fragments called functional groups instead of entire molecules.

This solves the problem of data unavailability because if the contribution of each functional group in a particular property is calculated once, then it remains the same regardless of the molecular structure in which it is used (*i.e.*, it is transferable in different molecules). As a result, simpler or complex molecular properties are calculated using GC models developed around databases of experimentally pre-determined property contributions for each functional group. GC methods provide predictions which are sufficiently accurate so that large molecular sets may be easily screened and few selected molecules of high performance in desired properties may then be validated using EoS, computational chemistry methods or experiments. Despite their obvious advantages they are challenged by the need to pre-specify a molecular structure (e.g., an ORC working fluid) in order to calculate its property values. This characteristic is also shared with EoS and computational chemistry methods and requires some prior knowledge regarding molecular structures that may lead to optimum ORC performance, otherwise the exhaustive examination of every possible molecule that exists is unavoidable in order to ensure the identification of a truly optimum ORC working fluid.

3.3. Optimization-Based CAMD of Pure Fluids

The above challenges are efficiently addressed by computer aided molecular design (CAMD) methods which combine the merits of GC methods with optimization algorithms. Papadopoulos *et al.* [4,5,37] proposed such an approach where an optimum molecule with desired properties is automatically identified based on the computational emulation of a molecular synthesis process (*i.e.*, the iterative transformation and evolution of an initial structure using different combinations of functional groups). An optimization algorithm guides the synthesis towards optimum structures, using properties as performance measures that reflect on molecular or process characteristics. The combination of GC methods with optimization also proves useful when a pre-specified database of molecules exists and requires fast screening to efficiently identify highly-performing options. CAMD approaches cover a very wide range of potentially optimum structures, support the identification of either novel molecular structures or conventional but previously overlooked, optimum molecules and rely on robust and systematic algorithms. Properties may be calculated by simpler GC models capturing the molecular chemistry effects on major ORC operating characteristics. EoS models may also be used (in combination with GC representations or not) to directly link molecular structure with ORC process economic and operating performance. Figure 3 illustrates the algorithmic steps involved in the optimization-based CAMD approach used in Papadopoulos *et al.* [4,5]:

- The selection of several functional groups from a database enables the generation of a molecule that is tested in terms of chemical feasibility.
- The desired properties of any feasible molecule are subsequently calculated based on the contribution of each functional group in the molecule.
- Several of these properties are used as a measure of molecular performance, *i.e.*, as objective functions in the employed optimization algorithm. The employed properties may directly reflect molecular characteristics or ORC process features.
- The optimization is then used to assess the performance based on specific algorithmic criteria and to inflict alterations in the molecular structure using functional groups available in the database, in order to generate a new molecule.

- This iterative procedure continues until a molecule with the optimum performance is identified, based on algorithmic termination criteria that ensure optimality.

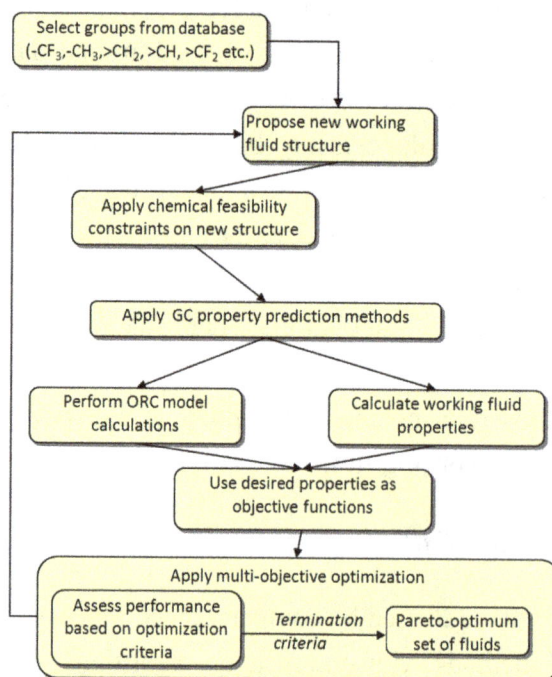

Figure 3. Algorithmic steps involved in the optimization-based CAMD approach used in Papadopoulos *et al.* [4,5].

Papadopoulos *et al.* [4,5,37] consider numerous molecular and ORC process-related properties as performance criteria in an approach which first employs CAMD to design an inclusive set of optimum working fluid candidates and then introduces several of them into ORC process simulations to select few that exhibit favorable process performance. At the CAMD stage properties calculated directly as a result of the working fluid structure involve density, latent heat of vaporization, liquid heat capacity, viscosity, thermal conductivity, melting point temperature, toxicity and flammability. A GC approach has been utilized for their calculation such as the one proposed by Hukkerikar *et al.* [27]. These properties reflect the effects of molecular chemistries on different desired ORC operating and design characteristics. For example, fluids of high density enable equipment of lower volume, fluids of low viscosity enhance the heat transfer hence requiring heat exchangers of lower area, the fluid heat capacity and enthalpy of vaporization have different effects on phase-change and superheating with impacts again on the heat exchanger sizes and cooling loads and so forth. All these properties are considered as objective functions in a multi-objective optimization problem formulation which is solved using Simulated Annealing and results in an inclusive set of Pareto optimum molecules. The development of a Pareto front enables the identification of useful trade-offs among the properties considered as objective functions, while molecules are designed to simultaneously optimize all properties. From a mathematical perspective, in this front no working fluid is of higher performance than the others simultaneously in all properties, but at least one of the properties of a working fluid is better than the same property of another fluid. At the same time, working fluids with worse performance than others in all properties are eliminated and steered clear of the non-dominated set [18,19,38]. The resulting candidate molecules in the Pareto front are then qualitatively evaluated based on their ozone depletion and global warming potentials considering structural rules from the literature. Selected molecules are introduced into ORC optimization in a basic system configuration. The aim of the optimization of the ORC process is to identify the heat exchange areas required in the vaporizer and the condenser that enable maximum energy recovery with minimum capital cost. The main findings that also illustrate the benefits of the proposed approach are the following:

- Several designed fluids are known chemicals, documented in the online NIST (www.nist.gov) or other databases, indicating the ability of CAMD to identify fluids which are readily available for utilization.
- Despite their public or commercial availability, many of the fluids obtained from CAMD have not been previously considered for ORC applications, indicating the ability of the method to point towards new design directions, overlooked by trial-and-error methods.
- The fluid 3,3,3-trifluoropropene only differs by a single fluorine atom from 2,3,3,3-tetrafluoropropene which has been commercialized in recent years by an international company [39] as an ORC working fluid, highlighting opportunities to quickly investigate other options which are very similar to the proposed designs.
- The fluid hexafluoropropane also obtained from CAMD has been mentioned in patents [40,41] as an ORC mixture component.
- Several unconventional and possibly novel working fluid structures were also identified combining ether and amine functional groups in fluorinated carbon chains. These groups were later shown to result in high ORC thermal efficiency in a study based on molecular thermodynamics [42] which accounted for the results of Papadopoulos et al. [4], among other fluids.

An optimization-based CAMD approach was also proposed by Palma-Flores et al. [43] which has similarities and differences with the work of Papadopoulos et al. [4,5,37] (overview of main points in Table 1). Palma-Flores et al. [43] solve an optimization-based CAMD problem which exploits group contribution methods for prediction of properties and also considers the feasibility of the molecular structures through appropriate constraints. Unlike Papadopoulos et al. [4,5] who employ Simulated Annealing as the optimization algorithm, Palma-Flores et al. [43] employ a Mixed Integer Non Linear Programming (MINLP) model which is solved with a deterministic optimization solver, namely DICOPT. Palma-Flores et al. [43] also solve the problem in two stages; first working fluids are designed using CAMD, while the resulting fluids are then compared in terms of ORC performance using three different process configurations. The CAMD stage is implemented 4 times using different objective functions and resulting in 32 working fluids which are further investigated in the second stage. The authors consider a more extensive set of functional groups than Papadopoulos et al. [4,5,37] including different aromatic and halogen options. Papadopoulos et al. [4,5,37] excluded these options due to issues with toxicity, ozone depletion and global warming. The findings of Palma-Flores et al. [43] seem to justify their exclusion. The objective functions include different combinations of working fluid liquid heat capacity, latent heat of vaporization and Gibbs free energy of formation. The first two properties are associated with heating, cooling and phase change operations in the cycle, while the latter is associated with the stability of the designed working fluids. Upper and lower bounds are implemented for the properties used as objective functions, while bounds are imposed on additional properties including critical pressure and temperature, normal boiling point and fusion temperatures. All bounds are obtained by investigation of the corresponding properties of some very common fluids previously utilized in ORCs. The authors note that some of the designed compounds have been previously considered in research literature but not as ORC working fluids. The authors use their GC models developed from literature sources to predict the working fluid properties hence they select several of them to compare their own predictions with results obtained from the ASPEN software. The observed deviations are mostly less than 5%, although in few occasions larger deviations are also observed. The designed fluids are then introduced in ORC process simulations considering three different process configurations; the basic ORC configuration, an ORC with an internal heat exchanger for heat recovery and an ORC with turbine bleeding and a direct contact heater. The fluids are evaluated considering their thermal efficiency in the different systems.

Lampe et al. [44,45] proposed an optimization-based method for the design of optimum ORC working fluids, namely the continuous molecular targeting (CoMT-CAMD) method (Table 1). Working fluids are designed based on a molecular model which allows the use of physical molecular characteristics as continuous decision parameters in the optimization problem. The molecular model takes the form of the perturbed chain statistical associating fluid theory (PC-SAFT) EoS which considers molecules as chains of spherical segments that interact through van der Waals interactions, hydrogen bonds, and polar interactions. The parameters considered in this work are the segment number and diameter as well as the van der Waals attraction between segments. This physical representation of the working fluid is used to calculate the residual Helmholtz energy which allows the calculation of the vapor-liquid equilibria in an ORC model. In this respect, the use of an EoS allows the direct employment of an ORC process model in fluid design and hence the utilization of a process-related objective function (e.g., ORC power output etc.). The resulting working fluids are represented by the optimum values of the segment number, diameter and the van der Waals attraction between segments, while Papadopoulos et al. [4,5,37] and Palma-Flores et al. [43] obtain optimum molecular structures. The resulting working fluid is therefore hypothetical in the sense that it does not necessarily coincide with a real fluid or satisfy chemical constraints (e.g., zero free bonds etc.). The authors address this issue by postulating a mapping stage where the parameters of the optimum working fluid are compared with the parameters of real working fluids contained in a database. The proximity of the optimum working fluid with the database fluids is evaluated based on the expected loss in ORC.

It is worth noting here that an approach for working fluid design and selection which shares similar features to Lampe et al. [44,45] has been recently proposed by Roskosch and Atakan [46]. The authors perform a reverse engineering design of the working fluid and a heat pump process (which has similarities with ORC) using a cubic EoS. Fluids are represented continuously in the optimization problem through critical temperature and pressure, acentric factor and liquid heat capacity. The problem is solved using non-linear programming (NLP). The resulting optimum solution is then identified based on its proximity to fluids available in a database. Additional criteria such as pressure limits, coefficient of performance and safety are also considered for the selection of the final fluids from the database. CAMD-based approaches addressing the design of refrigerant fluids and/or systems (which also have some similarities with ORC) have also been proposed by Samudra and Sahinidis [47], Sahinidis et al. [48], Duvedi and Achenie [49] using deterministic MINLP-based formulations and Marcoulaki and Kokossis [50] using Simulated Annealing.

Table 1. Main points in existing methods for the optimum design of pure ORC working fluids.

Main Points	Papadopoulos et al. [4,5,37] [a]	Palma-Flores et al. [43] [c]	Lampe et al. [44,45] [b]
Implemented stages	*Stage 1*: CAMD optimizing molecular structure. *Stage 2*: Evaluation of optimum molecules in ORC process optimization.	*Stage 1*: CAMD optimizing molecular structure. *Stage 2*: Evaluation of optimum molecules in ORC process simulation.	*Stage 1*: CoMT-CAMD optimizing molecular parameters and ORC process. *Stage 2*: Mapping of optimum molecular parameter values in molecular structures of existing molecules.
Property prediction method	GC + EoS; (e.g., standard cubic)	GC + EoS; (e.g., standard cubic)	PC-SAFT + QSPR (for ideal heat capacity)
Working fluid optimization parameters (*Stage 1*)	Functional groups (discrete, result in optimum structure)	Functional groups (discrete, result in optimum structure)	Segment number, diameter and van der Waals interactions (continuous, result in optimum values)
Optimization approach (*Stage 1*)	Multi-objective optimization, Simulated Annealing	Single objective optimization, MINLP solver	Single objective optimization, NLP solver
Working fluid optimization criteria (*Stage 1*)	Density, Enthalpy of vaporization, Liquid heat capacity, Viscosity, Thermal conductivity, Toxicity, Flammability, Melting point temperature, Critical temperature, Ozone depletion potential (qualitative), Global warming potential (qualitative).	Enthalpy of vaporization, Liquid heat capacity, ratio of the two, weighted sum of the two and the standard Gibbs energy of formation of an ideal gas.	ORC net power output.
Optimization criteria (*Stage 2*)	Unified index considering maximization of power output revenues and minimization of capital costs (vaporizer and condenser areas).	ORC thermal efficiency.	Expected loss in process performance of optimum (theoretical fluid) compared to real fluids in a database.
Identified fluids [a]	• CF_3-CH_2-CF_3 (Hexafluoropropane-R236fa) • CF_3-CH=CH_2 (Trifluoropropene-R1243) • CH_3-CH_2-CH_3 (Propane) • CH_3-O-NH-CH_3 • NH_2-CH_2-O-CH_3 • $HCOOCH_3$ • FCH_2-O-O-CH_2F • CH_3-O-O-CH_3	• CH_3-O-$N(OH)$-CH_3 • NH_2-O-CH_2-F • CH_3-CH_2-COO-CH_2-F • Cl-COO-CH_2-CH_3 • CH_3-O-O-$N(F)$-OH	• CF_3-CHF-CF_3 (Heptafluoropropane-R227ea) • CF_3-CH=CH_2 (Trifluoropropene-R1243) • CH_3-CH_2-CH_3 (Propane)

Heat source temperatures: [a] 90 °C, [b] 120 °C, Evaporator working fluid outlet temperature: [c] 190 °C.

3.4. Optimization-based CAMD of Mixtures

The design of mixtures is a considerably more challenging problem than the design of pure fluids; it requires the determination of (a) the optimum number of working fluids in the mixture, (b) the optimum mixture composition (*i.e.*, the structure of each mixture component) and (c) the optimum mixture concentration (*i.e.*, the amount of each component in the mixture). Papadopoulos *et al.* [18,19] proposed for the first time the design of binary ORC working fluid mixtures through a novel, optimization-based CAMD approach which may also be used for the design of mixtures in other applications. The proposed approach involves two main stages which are illustrated in Figure 4. The first stage aims to explore and identify the highest possible economic, operating, environmental and safety performance limits of a wide set of mixtures in an ORC system. This is approached in Stage 1 by searching for chemically feasible fluid structures only for one of the two components (*i.e.*, the 1st) of a binary mixture, while emulating the mixture behavior of the 2nd component within a much wider structural design space by lifting the chemical feasibility constraints. Note that in each stage the proposed approach enables the simultaneous mixture and ORC design. The two stages interact to help improve the performance of the obtained solutions. The proposed approach builds on the previous work of Papadopoulos *et al.* [4,5,37] for CAMD-based design of pure fluids hence the identification of multiple optimum mixture candidates is again accomplished through a multi-objective formulation of the CAMD-optimization problem, treating multiple ORC performance measures simultaneously and resulting in a comprehensive Pareto front revealing useful structural and property trade-offs among mixture components. Stage 2 serves to determine the optimum and chemically feasible structure of the 2nd component for each one of the feasible fluids (1st components) already obtained in Stage 1, together with the optimum mixture concentration. In Stage 2, the mixture performance limits identified in the previous stage are used as a reference point to efficiently avoid sub-optimal choices. The design of binary mixtures could in principle be approached directly in Stage 1 (*i.e.*, without the need for a second stage) by implementing chemical feasibility constraints on both new fluid structures. However, this may require increased computational effort, especially if such an approach is extended to mixtures of more than two components. Instead, the effort is reduced in the proposed approach as the user is allowed to review, interpret and analyze the rich intermediate insights generated by the multi-objective optimization approach prior to exploiting meaningful conclusions between design stages. Optimum solutions are identified in a Pareto sense, enabling the exploitation of the often conflicting design objectives. Some of the resulting mixtures are shown in Table 2, containing fluids that also favor the ORC performance even when they are used as pure fluids.

An approach addressing the optimization of working fluid mixtures for ORC is presented in Molina-Thierry and Flores-Tlacuahuac [51]. The number of working fluids participating in the mixture, the type of working fluids that form the mixture and the mixture concentration are optimized together with the ORC operating conditions. The working fluids that are used to perform mixture combinations are selected from a pre-specified set of three, eleven or six pure fluids in the performed case studies. This is different to Papadopoulos *et al.* [18,19] who identify the optimum structure of both working fluids participating in binary mixtures, simultaneously with the ORC operating conditions and without having a set of pre-specified options.

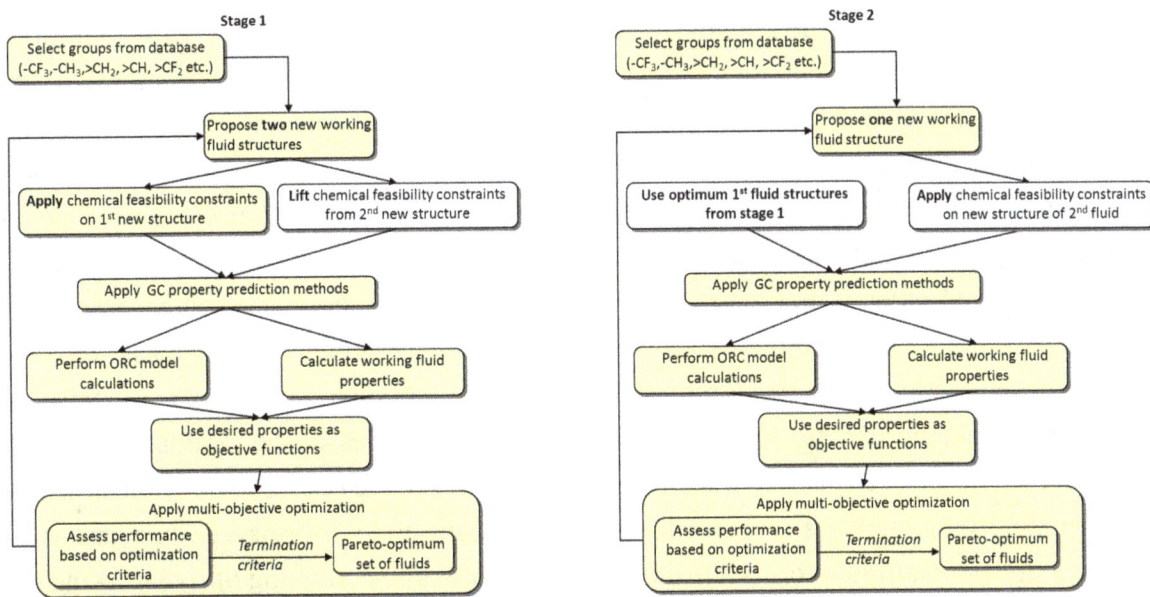

Figure 4. Algorithmic steps involved in the optimization-based CAMD approach for simultaneous mixture and ORC design used in Papadopoulos *et al.* [18,19].

Table 2. Main points in methods for optimum design and selection of ORC working fluid mixtures.

Main Points	Papadopoulos *et al.* [18,19] [d]	Molina-Thierry and Flores-Tlacuahuac, [51] [e]
Implemented stages	*Stage 1*: CAMD optimizing 1st and theoretical 2nd structure, concentration and ORC process. *Stage 2*: CAMD optimizing 1st and 2nd structure, concentration and ORC process.	Optimizing number and type of working fluids in mixture (generated from a pre-specified set of pure fluids), mixture concentration and ORC process.
Property prediction method	GC + EoS (e.g., standard cubic)	GC + EoS (e.g., standard cubic)
Working fluid optimization parameters (*Stage 1*)	Functional groups (discrete, result in optimum structure), concentration (continuous)	Preselected set of pure fluids used to form mixture combinations (discrete), concentration (continuous)
Optimization approach (*Stage 1*)	Multi-objective optimization, Simulated Annealing	Single objective optimization, testing of several objectives, MINLP solver
Working fluid and process optimization criteria (*Stage 1*)	Exergetic efficiency, thermal efficiency, flammability of each fluid, mixture maximum and minimum flash points (constraint), mixture azetropic concentration (constraint).	Change of the enthalpy of vaporization at the cycle high pressure level, specific net work output, first and second law efficiency, area in between profiles (temperature-enthalpy) of the working fluid and heat source or sink on the heat exchanger
Optimization criteria (*Stage 2*)	Same as *Stage 1*	Not applicable
Uncertainty in mixture selection	Considered through a systematic non-linear sensitivity analysis approach	-
Identified fluids	• CF_3-CH_2-CH_3/FCH_2-O-$(CH_2)_2$-CH_3 • CF_3-CH_2-CF_3/FCH_2-O-CH_2-CF_3 • (CH_3)-C/FCH_2-O-C-$(CH_3)_3$	• RC318 (refrigerant)-*n*-Pentane (case B) • R245ca- *n*-Pentane (case B) • FC4-1-12-*n*-Butane (case C)

Heat source temperatures: [d] 90 °C, [e] 90–150 °C.

The authors present details on the employed ORC model which involves vapor-liquid equilibrium and other calculations using models similar to Papadopoulos *et al.* [18,19]. They employ a single objective formulation, trying five different simple or complex objective functions during optimization and considering three different case studies. The optimization results in the same binary mixture components in the first case study, with changes observed only in the mixture concentration depending on the employed objective function. In the second case study the authors use the first law efficiency as an objective function and solve the optimization problem for several different heat source temperatures. The mixture compositions and concentrations change significantly. Although in some cases the results indicate mixtures consisting of four components, only two of them are in significantly high concentration, the remaining two are traces. The third case study addresses the optimization problem independently for two objective functions indicating that the choice of the objective function affects both the optimum mixture characteristics (composition, concentration) and performance. One case finds a binary mixture as the optimum solution, while the other finds a mixture consisting of five working fluids in significant concentrations. The obtained mixtures are not comparable with Papadopoulos *et al.* [18,19] because the latter used different functional groups. The authors note that in the future they will also consider uncertainty as well as process control (discussed in the subsequent sections).

3.5. Uncertainty in Predictions

The use of predictive models for the calculation of working fluid or ORC process properties involves uncertainty regarding the accuracy of the obtained predictions. Uncertainty is mainly observed in the employed GC, thermodynamic or process models and results in over- or under- estimation of the predicted thermodynamic or process behavior of the investigated or designed fluids. The use of different prediction models or input parameters for the calculation of the same property may result in values that deviate. While deviations may be significant for particular fluids, others may exhibit similar property values regardless of the employed property prediction model. Lampe *et al.* [45] illustrate the impact of the working fluid structure on the ORC net power output through their continuous molecular representation using the geometric and interaction parameters. The reported three-dimensional diagram indicates areas with very steep non-linear changes as well as areas with smoother changes. In other words, the sensitivity of the employed model under the influence of different fluids may vary significantly. In the case that experimental measurements are available it is possible to determine the accuracy of the predictive models and hence ensure that by accounting for predictive inaccuracies the designed or selected fluids represent realistic ORC performance options. However, experimental measurements exist for very few working fluids under very specific conditions. It is therefore necessary to utilize a systematic method which enables the validation of the obtained predictions with respect to their expected accuracy, regardless of the availability of experimental measurements or the predictive capability of the available models.

Papadopoulos *et al.* [18,19] proposed a sensitivity analysis approach which facilitates the identification of parameters with high influence in the overall working fluid-ORC system performance, the quantification of the overall system sensitivity with respect to these parameters and the incorporation of sensitivity metrics during the decision-making involved in the optimum working fluid selection. The proposed method identifies ORC process performance areas that present steeper or smoother changes for different fluids under the simultaneous influence of multiple different parameters for each fluid and determines the parameters with the highest influence in the changes. The method was implemented in the selection of optimum ORC working fluid mixtures and may be also be used for the selection of pure working fluids. It is based on the development of a sensitivity matrix which incorporates the derivatives of the ORC performance measures (e.g., thermal or exergetic efficiency *etc.*) with respect to model parameters and constitutes a measure of the variation of the employed model under the influence of infinitesimal changes imposed on model parameters. The sensitivity matrix is decomposed into major directions of variability to identify the largest in magnitude eigenvector. This represents the dominant direction of variability for the system, causing the largest change in the performance measures. The entries in the dominant eigenvector determine the major direction of variability in the multiparametric space and indicate the impact of each parameter in this direction. Having identified this direction it is not necessary to explore all directions of variability (*i.e.*, combinations of parameters) arbitrarily hence reducing the dimensionality of the sensitivity analysis problem. The dominant eigenvector is then exploited in a sensitivity index which accounts for all performance indices simultaneously within a wide variation range explored also through an appropriate parameter. In this respect, the optimum working fluid mixtures which resulted from CAMD were also evaluated in terms of the accuracy in the performance predictions. Some mixtures that exhibited high ORC performance seemed to be very sensitive in changes in certain model input parameters; in case that these parameter values are not sufficiently accurate the predicted performance will drop significantly. The predicted performance of several other mixtures that exhibited low sensitivity would not be affected even if the model input parameter values were less accurate.

3.6. Simultaneous vs. Integrated Design Approaches

The reviewed cases reveal that the design and/or selection of working fluids follows two major approaches:

- An approach that supports the *simultaneous* working fluid and ORC design and/or selection (Palma-Flores *et al.* [43]; Lampe *et al.* [44,45]; Papadopoulos *et al.* [18,19]; Molina-Thierry and Flores-Tlacuahuac [51]).
- An approach that supports the *integrated* working fluid and ORC design and/or selection (Papadopoulos *et al.* [4,5,37]).

In most of the above cases the pure or mixed working fluid structure(s) are designed using a CAMD approach. There is also an option of determining an optimum pure working fluid or the mixture composition and concentration from a pre-specified list of working fluids (of known structures) or their combinations. This may be done either through a *simultaneous* or an *integrated* approach.

- The meaning of *simultaneous* is that decisions regarding the working fluid structure, composition or concentration (in case of mixtures) are taken within the same optimization algorithm that identifies the optimum ORC operating and/or sizing characteristics. The advantage of a *simultaneous* approach is that the working fluid and ORC interactions are accounted for together and drive the optimization search to identify an optimum solution. This is reasonable because a working fluid is an inherent component of the ORC system in which it is utilized. However, simultaneous approaches may suffer from combinatorial complexity if the design options in both the working fluid and ORC sides are extensive. The relevant reviewed works incorporate a basic ORC structure into working fluid design in order to identify fluids directly based on their impact in the ORC process. It would be ideal to also consider the structural characteristics of the ORC (e.g., recuperation, pressure levels *etc.*) simultaneously with working fluid design but this would lead to an intractable optimization problem due to the vast number of potential options and the non-linearities of the employed models. For example, Palma-Flores *et al.* [43] consider more complex ORC structures in simulations performed after the optimum working fluids were identified.

- The meaning of an *integrated* design approach is that working fluids are first designed using fluid properties as objective functions and the obtained fluid(s) are then introduced in a full ORC model simulation or optimization, as in the case of pure fluid design presented in Papadopoulos *et al.* [4,5]. *Integrated* design approaches evolved from the need to decompose the CAMD and process design problems so that an extensive search space (e.g., working fluid and process structures, operating parameters *etc.*) may be considered within reasonable computational efficiency. The key to the efficient integration of a CAMD and a process design stage is to identify the working fluid(s) that will subsequently facilitate high performance in ORC optimization, while avoiding the premature exclusion of useful fluid options. In a broader sense *simultaneous* approaches could become part of the problem decomposition rationale employed in *integrated* approaches. This is because at some point optimum working fluids and process characteristics obtained from the *simultaneous* design stage will have to be transferred to a subsequent, independent design stage to perform optimizations either using more detailed and realistic models or exploring a much wider design space. *Integrated* approaches address the problem of obtaining and transferring useful and inclusive design information in the subsequent stage.

Table 3. Merits and shortcomings of considered methods and objective functions in fluids design.

	Merits	Shortcomings
Fluid selection from a pre-specified list	Few options to investigate in an optimum selection problem of reduced size, ORC model may be used, selection may also be based only on molecular properties, commercially available fluids may be used directly	The consideration of few options limits the search, arbitrarily excluded working fluids may be the ones that maximize ORC performance, novel working fluids may not be identified
Simultaneous design methods	Directly link molecular characteristics with ORC process performance, novel working fluids may be identified	Require an ORC model, which may however reduce computational efficiency if modeling rigor is increased
Integrated design methods	Enable the consideration of a more extensive design space and/or more detailed models in each design stage, maintain manageable computational effort, novel working fluids may be identified	Require efficient integration between stages to avoid excluding important designs early on in the search
Molecular properties as objectives	Easy to calculate and implement through GC methods, no need for an ORC model, appropriate for CAMD and multi-objective formulations	Indirectly reflect on ORC process performance characteristics, not appropriate for use in single-objective formulations

Table 3. *Cont.*

	Merits	Shortcomings
ORC process performance indices as objectives	Directly link CAMD with ORC performance, provide associations with cost/profit characteristics (e.g., net generated work) or direct use of them, support more realistic solutions when more detailed models are used	Detailed ORC models may impact on computational efficiency, associations to cost/profit through less detailed models may be limited to specific ORC characteristics, overlooking other important costs, depending on the model detail
Single-objective formulations	Easy to implement, result in a single optimum solution	The use of different properties as the objective function is likely to result in different optimum working fluids
Multi-objective formulations	Handle multiple and often conflicting objectives simultaneously, result in a rich set of working fluids, appropriate for molecular properties	More difficult to implement than single-objective formulation

Table 3 provides an overview of merits and shortcomings. Note that *simultaneous* and *integrated* approaches are not related to the type of the employed optimization algorithm which may either be stochastic such as Simulated Annealing *etc.* (Papadopoulos and Linke [52]) or deterministic such as NLP- or MINLP-based algorithms *etc.* (Cavazzuti [53]) in both cases. Optimization algorithms are discussed in the next section.

3.7. Single- vs. Multi-objective Optimization and Types of Objective Functions

Regardless of the employed approach, molecular properties are often used as working fluid screening or design criteria because they indirectly reflect on ORC process performance. For example, high working fluid density may enable a reduction in the required fluid amount hence equipment of lower size may be used. High thermal conductivity enables heat exchangers of lower areas and so forth. In *single-objective* optimization formulations the use of a molecular property as objective function may lead to optimum working fluids which are different depending on the selected property. They may also be different to the results obtained from a simultaneous approach. These challenges are best addressed by the use of a *multi-objective* optimization formulation in the working fluid design stage (Table 3). In such a case:

- There is no need to select one of the properties as an objective function in order to guide CAMD into the identification of a single optimum molecule, while there is no need to set upper and lower bounds (which are often not known a priori) in the remaining properties which are used as constraints. This is very important because there are many working fluid properties that may be considered as objective functions and a *multi-objective* formulation does not have limitations on how many may be included.
- Such an approach results in a Pareto front which consists of many working fluids, instead of one obtained in a *single-objective* case. The Pareto fluids represent multi-fold and rich trade-offs in the entire ORC performance spectrum. They can be incorporated as discrete options in a subsequent comprehensive ORC design stage (see next section) where the optimization and computational efficiency remain manageable (Papadopoulos and Linke [16]).

As shown in Table 1 the optimum working fluids obtained from the integrated design approach of Papadopoulos et al. [4,5] are very similar to those obtained by the simultaneous approach of Lampe et al. [44,45]. Papadopoulos et al. [4,5] were able to capture such working fluids without the use of an ORC model in the course of working fluid CAMD (hence the computations were fast) due to the use of multiple property objective functions. Note that the results are comparable because the heat source temperatures are quite close and the fluids are pure. Papadopoulos et al. [37] reported that for heat source temperatures between 70–90 °C the rank ordering of working fluids in terms of ORC performance remained the same. Similar findings with respect to the selected fluids have been previously reported by Papadopoulos and Linke [38] who compared a simultaneous and an integrated *multi-objective* CAMD approach in a different application (*i.e.*, solvents for industrial separations).

The type of properties that could be used as objective functions in a single- or multi-objective CAMD working fluid formulation have been thoroughly discussed in Stijepovic et al. [17], while insights have also been discussed in Papadopoulos et al. [4] and Palma-Flores et al. [43].

In the case of pure fluids:

- Stijepovic et al. [17] investigated the impact of different working fluid properties in the cycle thermal, exergetic efficiency and economics as a function of heat transfer areas and net generated work. It was found that high values of fluid compressibility factor and low values of saturated liquid molar volumes favor all three criteria. Fluids of high molecular weights favor thermal efficiency and of high isobaric heat capacities favor exergetic efficiency and economic performance, whereas low enthalpy of vaporization also favors the same criteria. Low critical pressure, high saturated liquid thermal conductivity and saturated gas volume favor economic performance.
- Palma-Flores et al. [43] report that the molecules resulting from minimization of liquid heat capacity and of a complex objective that combines a weighted sum of enthalpy of vaporization, liquid heat capacity and standard Gibbs energy of formation of an ideal gas result in higher thermal efficiency and work output.

In the case of mixtures:

- Papadopoulos et al. [18] finds that exergetic efficiency increases at a higher rate than thermal efficiency decreases, as the concentration moves from pure component to approximately equal amounts of components in the mixture. The use of a multi-objective approach appears helpful.
- Molina-Thierry and Flores-Tlacuahuac [51] find that the first law efficiency is the most appropriate objective to use in a single objective formulation.

Note that objective functions associated with costs are clearly useful but difficult to use when the goal is to design or screen for efficient working fluids. Even when an ORC model is used as part of working fluid design it is mainly based on a thermodynamic representation which indirectly associates cost with indices like net generated power. Papadopoulos et al. [4] note that about 90% of process costs are associated with heat exchangers. Although it would be desirable to calculate heat exchanger areas in the course of working fluid design, it is not practical mainly due to data limitations (e.g., heat transfer coefficients *etc.*). Furthermore, there are also numerous and complex economic performance indices which influence the optimum solution but are impractical to use in working fluid design due to limitations in the rigor of the employed model. Kasaš et al. [54] note that only those process models with sufficient levels of accuracy are suitable for generating proper optimal designs using the correct economic criterion. Such issues are discussed in the next section.

4. Optimization Approaches for Organic Rankine Cycles

4.1. Main Concepts and Computational Challenges

The careful selection of the working fluid is instrumental to the performance of the ORC as discussed above. Equally importantly, an ORC process configuration needs to be determined to enable efficient power generation for the given heat source(s). In addition, the cycle operating conditions need to be set such that the chosen performance criterion is maximized. The overall design problem requires discrete decisions on the selection of structural design alternatives such as alternative heat exchanger options or selection of the number of cycles to integrate as well as optimization of the continuous variables associated with operating conditions and equipment sizes. Systematic approaches can aid the exploration of the design options to guide the identification and selection of efficient overall designs. The computational challenges are similar to those observed in working fluid design. A very large number of structural and operating options need to be considered as part of the ORC flowsheet in order to identify an optimum system of high efficiency, whereas sufficiently accurate process models are also required so that the obtained designs are realistic. The three general model types that may be considered for the design of ORCs involve (Figure 5):

(a) Computational fluid dynamic (CFD) models for detailed equipment design [55].
(b) Process level models that incorporate thermodynamic calculations with equipment details for equipment sizing within the flowsheet.
(c) Thermodynamic cycle models which account for energy balances and phase change operations.

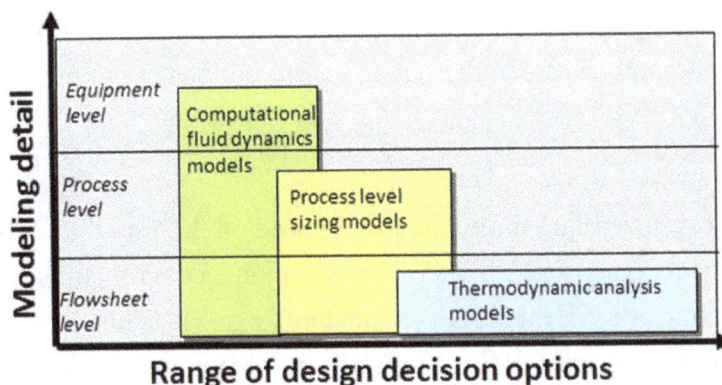

Figure 5. Equipment, process and flowsheet models with respect to modeling detail and range of design decision options which may be simulated at a reasonable computational effort.

The simultaneous utilization of representatives from all model types would be ideal but computationally impractical. Each model type may be used independently but not all types are suitable for the optimum design of ORC flowsheets. CFD models capture local heat and/or mass transfer phenomena within the equipment with great detail, but the computations are time consuming. Process level sizing models enable the consideration of economic performance measures to evaluate different flowsheet alternatives and allow the consideration of an increased range of decision options within reasonable computational efficiency. However, the use of such models in the course of working fluid design or selection (*i.e.*, with the molecular structure as an additional design parameter) may be limited by the lack of data (e.g., prediction models of heat transfer coefficients for different working fluids are quite complex [56]) and by the high combinatorial complexity of the design problem. Finally, thermodynamic analysis models enable the consideration of a much wider range of process and flowsheet design decision options, they have been used in the course of working fluid design as shown in the previous section and provide useful insights about different ORC flowsheet configurations prior to proceeding into a more rigorous evaluation.

The literature is abound with works that explore cycle operating conditions and structures largely through sensitivity analyses via repeat simulation studies to identify efficient settings for selected designs and given heat sources and sinks. The limitation of such contributions lies in the need for prior knowledge of an ORC configuration which may lead to good performance. Although empirical knowledge is very useful, the trial-and-error testing of different ORC configurations is likely to enable incremental performance improvement. Such contributions will not be reviewed here; instead, we will limit our focus to emerging systematic approaches to design efficient ORC systems. A number of such optimization-based approaches have been proposed in the last few years. The general ORC design optimization problem for such approaches can be stated as: *Given a heat source stream and ambient conditions, determine the optimal cycle configuration and design parameters that maximize ORC performance with respect to chosen performance criteria.* Notice that in this formulation it is not necessary to pre-specify an ORC structural or operating profile that will be optimized. Structural and operating ORC parameters may become decision variables in an optimization search which is guided toward the optimum solution by the chosen performance criteria through an algorithmic sequence. This does not eliminate the need for empirical knowledge which may be used to focus the design space into significant decision options or to interpret the design results in terms of their practical applicability. Earlier works focus at optimizing the design and/or operating parameters of ORC systems, whilst the most recent contributions attempt to consider alternative configurations in optimization approaches.

4.2. Reviewed Approaches

The reviewed works are organized into three categories based on the way that the cycle and working fluids are selected (Tables 4–6). Franco and Villani [57] were amongst the first to present an optimization scheme to help the identification of efficient design parameters for an ORC in a binary geothermal power plant. They propose to decompose the ORC optimization problem into three subsystems, the heat recovery cycle, the heat recovery exchanger and the cooling section. The three subsystems are evaluated in an overall iterative scheme where operating characteristics of the cycle are determined based on the optimization of the heat recovery and cooling system sizes. The performance measures are the first and second law efficiencies as well as the brine consumption from implementation of the system for a geothermal heat source. Six pre-selected working fluids are considered together with three ORC structures (supercritical, dual pressure level, ORC with superheater) which are all evaluated. Details on the implementation of the optimal search and convergence properties have not been provided.

With a different application focus, Salcedo *et al.* [58] propose a multi-objective optimization formulation for an integrated system of a solar ORC and a reverse osmosis desalination plant. The approach allows to explore the equipment sizes and operational conditions of a predefined system configuration and considers two objectives: Cost of water produced and the life cycle global warming potential. The resulting MINLP problem is solved by exhaustive enumeration using a deterministic nonlinear optimization algorithm. With respect to a standard ORC configuration, Wang *et al.* [59] propose an optimization approach to determine optimal design parameters using global search schemes in the form of meta-heuristics. The approach allows one to determine the optimal turbine inlet pressure and temperature as well as temperatures against heat sources and sinks. The ratio of net work to heat transfer as an indicator of cost is maximized as the design objective. A Genetic Algorithm is implemented to solve the nonlinear continuous optimization problem.

Table 4. Cycle operating/sizing parameters are optimized for different, pre-determined cycle structure and working fluid combinations.

Authors	Optimization Approach	Decision Parameters	Objective Functions
Franco and Villani [57]	Iterative hierarchical identification of optimum ORC size and operating parameters for each combination	Six pure working fluids, sizes of cooling systems and recovery heat exchanger, three cycle structures (supercritical, dual pressure level, ORC with superheater)	First, second law efficiency, brine consumption
Salcedo *et al.* [58]	MINLP problem, deterministic nonlinear algorithm	Equipment sizes, operating conditions, one cycle structure	Cost of water produced in desalination plant, global warming potential
Wang *et al.* [59–61]	Genetic algorithm [59,62], multi-objective genetic algorithm [60,61]	Turbine inlet pressure and temperature, temperatures against heat sources and sinks [59,60], geometry of ORC heat exchanger [61], one cycle structure	Net power output to total heat transfer area [59], exergetic efficiency and capital cost [60], pressure drop, heat transfer area [61]
Xi *et al.* [62]	Genetic Algorithm	Three regenerative cycle structures, six pure working fluids, operating parameters	Exergetic efficiency
Walraven *et al.* [63,64]	Deterministic NLP	Eight pure working fluids, tube diameters, baffle spacing in heat exchangers, two different cycle structures	Levelized cost of electricity [63], net present value [64]
Victor *et al.* [30]	Simulated Annealing	Composition of working fluid mixture, ORC, Kalina cycle	Thermal efficiency

Wang *et al.* [60] later proposed the solution of a problem considering two objectives simultaneously, the exergetic efficiency and the capital cost. The Pareto frontier is determined using a multi-objective Genetic Algorithm. Xi *et al.* [62] propose a Genetic Algorithm based approach for parameter optimization of regenerative ORC configurations to achieve maximum exergetic efficiency.

The application of the approach is illustrated with a study of three different cycle configurations and six different working fluids in an exhaustive search. At the level of equipment design, Wang *et al.* [61] optimize the geometry of ORC plate heat exchanger condensers using a multi-objective Genetic Algorithm to explore the nondominated solutions with respect to pressure drop and heat transfer area.

Walraven *et al.* [63] present a parameter optimization scheme for single and multi-pressure ORCs that takes into account the geometry of shell-and-tube heat exchangers for the heat recovery section as well as models for dry cooling. Multi-pressure ORCs are represented by flowsheets where the working fluid is compressed in two or more loops at different temperature levels. Dry cooling refers to air cooled condensers. In contrast to the more prominent global search algorithms based on meta-heuristics, the optimization problem is solved using a local deterministic NLP solver to determine the optimized ORC design parameters such as tube diameters and baffle spacing that yield the maximum Net Present Value of the system. Structural variants are again explored through repeat solutions of pre-selected configurations. Earlier, Walraven *et al.* [64] presented a similar optimization scheme that uses the Levelized Cost of Electricity as the objective function taking into consideration wet and air cooling. Wet cooling refers to a water-cooled tower.

Moving beyond pure component working fluids, Victor *et al.* [30] consider ORCs and Kalina cycles with working fluid mixtures and propose an optimization approach to determine the optimum working fluid composition that maximizes the thermal efficiency of the cycle. The optimization problem is implemented using the Simulated Annealing meta-heuristic to perform a global search of the solution space.

Gerber and Marechal [65] proposed a multi-period, multi-objective optimization approach to determine optimal configurations for geothermal systems while accounting for seasonality. Multi-period optimization is used to enable the incorporation of parameter variation in the design procedure. The optimization is based on the expected value of the objective function for a given parameter variability. Usually, parameter variation is imposed through the consideration of multiple parameter realizations taken from the selected parameter space as discrete instances of the plant (periods) [66]. The overall approach employs an evolutionary algorithm across the multiple periods considered and draws on process integration approaches and the solution of single period mixed-integer linear programming (MILP) problems to determine the optimal configurations with respect to objective functions including the investment and operating cost and the exergetic and energetic efficiency. The problem involves the selection or combination of different energy technologies for the exploitation of geothermal fields at different depths. Among other technologies, two ORCs are considered, one single-loop and one with intermediate draw-off. The decision parameters for the ORCs involve the evaporation and saturation temperature in both cases as well as draw-off split fraction and condensation temperature in the second case.

Taking into account structural ORC design decisions, Pierobon *et al.* [67] propose an optimization approach to simultaneously explore design parameters and structural alternatives for ORC configurations and single working fluid options from a predefined set of candidates. The approach implements a Genetic Algorithm to simultaneously explore the design space for three objective functions: The Net present Value, the total system volume and thermal efficiency. The approach simultaneously determines the best working fluid from a predefined set, sizes the heat exchangers, and determines the temperature and pressure settings of the cycle. Larsen *et al.* [68] present a variation of the work to simultaneously explore alternatives for ORC configurations in terms of internal heat recovery and superheating options. A data set of 109 working fluids is screened before the optimization based on thermodynamic and hazard criteria. Few working fluids are optimized simultaneously with the ORC structural and operating options. Both works study waste heat recovery on an offshore platform to illustrate the approach.

Most recently, Clarke and McLeskey [69] have proposed a multi-objective optimization approach for ORC system design. Their approach allows to simultaneously consider two objective functions, the specific work output of the ORC and the specific heat exchanger area, and six decision variables: The choice of alternative working fluids out of a predefined set of 17 candidates, the evaporation temperature, the minimum approach temperature, the effectiveness of the superheater, the effectiveness of the recuperator and the temperature difference in the condenser. The Pareto front is developed using a Particle Swarm optimization algorithm which has been implemented for geothermal power generation. The benefit from the optimization tool in exploring the design options is highlighted.

Table 5. Optimum cycle structure, operating/sizing parameters and working fluids are selected simultaneously from a pre-determined set of options.

Authors	Optimization Approach	Decision Parameters	Objective Functions
Gerber and Marechal [65]	Multi-period, multi-objective, evolutionary algorithm across multiple periods, MILP in single periods (cycle structure and operation)	Two cycle structures (single-loop, intermediate draw-off), evaporation and saturation temperature (both structures), draw-off split fraction and condensation temperature (2nd structure)	Investment and operating cost, exergetic and energetic efficiency
Pierobon et al. [67]	Genetic algorithm (cycle operation and working fluid)	Five pure working fluids, size of heat exchangers, turbine inlet pressure and temperature, the condensing temperature, pinch points, superheating temperature difference, target velocities in heat exchangers	Thermal efficiency, total system volume, net present value
Larsen et al. [68]	Genetic algorithm (cycle structure, operation and working fluid)	Set of working fluids, structures with recuperation and/or superheating	Thermal efficiency
Clarke and McLeskey [69]	Multi-objective particle swarm (cycle operation and working fluid)	Seventeen working fluids, evaporation temperature, minimum approach temperature, effectiveness of superheater and recuperator, temperature difference in condenser	Specific work output, specific heat exchanger area

The methods presented thus far aim at the optimization of fixed ORC configurations with a pure working fluid. In a recent attempt to broaden the search towards including a broader set of structural design candidates, Stijepovic et al. [70] propose a method for the optimal design of multi-pressure ORCs to generate power form a single heat source stream. The approach draws on the Exergy Composite Approach by Linnhoff and Dhole [71] to formulate an optimization problem that is repeatedly solved to determine the ORC configuration and its optimal operating conditions with minimum exergy loss and maximum work output. The work considers both induction and expansion turbines. The presented results highlight significant performance improvements of the developed multi-pressure ORC configurations over the standard single-pressure ORC configuration. Toffolo [72] proposes an optimization approach to determine optimal configurations and design parameters for ORCs that absorb and release heat at different temperatures. The approach combines a Genetic Algorithm (GA) to screen configurations with a sequential quadratic programming (SQP) approach to determine design parameters. It is based on a generic flowsheet representation which may be used to determine different topologies including different numbers of pressure and expansion levels as well as heat exchange operations.

Table 6. Optimum cycle structure and operating/sizing parameters evolve during the optimization search; the optimum cycle structure is not pre-determined but results from optimization.

Authors	Optimization Approach	Decision Parameters	Objective Functions
Stijepovic *et al.* [70]	Iterative addition of pressure loops to optimize an evolving structure, deterministic NLP inside each loop to optimize the operating parameters	Number of pressure loops, working fluid flowrates, saturation temperatures, evaporator outlet temperatures per loop, two expandable multi-pressure ORC configurations, type of turbine (induction, expansion)	Exergy loss, work output
Toffolo [72]	Genetic algorithm to synthesize structure, sequential quadratic programming to optimize objective function	Number and configuration of pressure loops, expansion and heat exchange stages	Net generated electrical power

4.3. ORC Process Structure Classes and Types

In summary, the past five years have seen the emergence of optimal design approaches for ORCs. Based on the performed review these approaches may be broken down into three general classes:

- *Fixed flowsheet approaches*: Those that focus on parameter optimization as well as on addressing equipment design decisions for a pre-specified ORC flowsheet.
- *Flowsheet selection approaches*: Those that consider cycle operating and equipment design for different pre-specified flowsheets. In such cases the pre-specified flowsheet structures may be decision parameters in the optimization or each structure may be optimized separately, one-by-one in terms of operating and equipment characteristics.
- *Flowsheet design approaches*: Those that have broadened the scope towards the inclusion of structural design decisions within the cycle. In such cases the flowsheet structure is not entirely pre-specified but a flexible structure gradually evolves into different configurations and the optimum flowsheet results from the optimization, together with operating and equipment characteristics.

This trend is expected to continue to yield systematic design approaches that can simultaneously consider the design parameters together with structural design alternatives associated with multiple working fluids, multiple integrated cycles and multiple pressure levels. The availability of such methods will be instrumental to the quick determination of optimal ORC-based power generation schemes for any given heat source and sink. A recent review by Lecompte *et al.* [73] provides an elaboration of several ORC structures studied in literature, based on the goals that they intend to address:

- Structures that intend to decrease irreversibility and match the temperature profiles between heat source and the working fluid involve transcritical cycles, trilateral cycles, cycles with zeotropic mixtures as working fluids, cycles with multiple evaporation pressures, organic flash cycles and cascade cycles.
- Structures that intend to increase thermal efficiency by maximizing the mean temperature difference between heat addition and heat rejection involve cycles with the addition of a recuperator, Regenerative cycles with turbine bleeding, cycles with reheaters and cycles with vapor injector.

4.4. Stochastic vs. Deterministic Optimization Methods

The key practical issues to be considered in the selection of an appropriate optimization method are the existence of integer and/or continuous design variables, of non-linearities (e.g., convexities or non-convexities) in the employed working fluid or ORC models and the quality of the obtained solutions (globally *vs.* locally optimum solutions). The two main existing categories are *deterministic* and *stochastic* optimization methods [53] (Table 7).

- *Deterministic* optimization methods exploit analytical properties (e.g., convexity and monotonicity) of the problem to generate a deterministic sequence of points converging to an (local or global) optimal solution [74]. They are often represented through variations of NLP (continuous variables) and MINLP (integer and continuous variables) problem formulations, although there are also several other problem classes [75]. They provide insights regarding the local [74] or global optimality of a solution through analytical mathematical conditions [75]. From a practical perspective they require a lower number of objective function evaluations to reach an optimum solution than *stochastic* methods and enable the identification of locally [53] or globally [75] optimum solutions in non-convex problems. Limitations of these methods involve the computationally intensive use of derivative transformations and difficulties in the initialization of simulations when complex models are considered. A fundamental issue of deterministic methods is to transcend local optimality [74] hence the development of mechanisms to prevent the convergence in local optima in highly non-convex problems is also a very active research field [76].
- *Stochastic* optimization methods, *i.e.*, methods for which the outcome is random, are particularly suited for problems that possess no known structure that can be exploited. These methods generally require little or no additional assumptions on the optimization problem [74]. The three main classes of stochastic methods are: Two-phase methods, random search methods, and random function methods [74]. The most well-known representatives of stochastic methods are Genetic Algorithms and Simulated Annealing [52,53] which are also called metaheuristics. Simulated Annealing is a typical representative of random search methods which is easily implementable, robust and applicable to a very general class of global optimization problems [74]. Metaheuristics usually emulate physical systems in order to explore the solution space of a given problem and identify the optimum solution through a series of probabilistic transformations. These methods do not suffer from the same limitations as the deterministic methods because their inherent mathematical operations are simple, their algorithmic mechanisms provide venues to target the globally optimal domain and discrete design parameters are handled easier. They can even be applied to ill-structured problems for which no efficient local search procedures exist [74]. However, the lack of these limitations is traded-off for convergence to a distribution of nearly optimal solutions and sometimes for long computational times required for the implementation of the stochastic runs. These characteristics are not necessarily shortcomings as the distribution of nearly optimal scenarios provides statistical guarantees for the quality of the solutions. These methods are very useful at early design stages when there is a vast number of discrete or continuous decision options to be investigated. The existence of multiple close-to-target optimum solutions provides valuable design insights into the problem which can be reviewed and analyzed by users prior to transferring

meaningful conclusions onto a subsequent stage where the design problem can be defined with considerably less uncertainty.

Table 7. Merits and shortcomings of considered methods and objective functions in process optimization.

	Merits	Shortcomings
Deterministic methods	Fewer function evaluations than stochastic methods to reach an optimum, analytical mathematical determination of local or global optimum	Intensive computations, difficult simulation initialization in non-convex models, mechanisms to avoid local optima is an active research field, require knowledge of analytical problem properties (e.g., convexity, monotonicity)
Stochastic methods	Suitable for early stage design with extensive and discretized design spaces, easier to implement than deterministic methods, no knowledge of optimization problem structure is required, rich design insights from close but different optimum solutions	Larger number of function evaluations to identify optimum solution, statistical assessment of solution optimality
Single-objective formulations	Easy to implement, result in a single optimum solution	Need for well-defined problems, use of different objective functions results in different designs, an appropriate objective function needs to be selected
Multi-objective formulations	Handle multiple and often conflicting objectives simultaneously, results in a rich set of finite designs representing important trade-offs	More difficult to implement than single-objective formulation
Thermodynamic objectives	Useful for early design stages using less rigorous process models	Appropriate objectives need to be selected and combined, indirect and approximate association with costs
Economic objectives	Support detailed and realistic designs when used with sufficiently detailed process models	More complex objectives than cost or profit may be needed, appropriate objectives should be selected based on optimization formulations and goals

In the reviewed works the preference for global search algorithms based on meta-heurists such as Genetic Algorithms, Simulated Annealing or Particle Swarm Optimization over deterministic optimization algorithms is noticeable. Approaches based on deterministic global optimization techniques [75] remain yet to be implemented for ORC design problems. Among different software packages, the GAMS software (www.gams.com) includes several deterministic optimization solvers, while MATLAB (www.mathworks.com) includes both deterministic and stochastic solvers.

4.5. Single- vs. Multi-Objective Optimization and Types of Objective Functions

Another observation is the use of *multi-objective* optimization approaches which allows the simultaneous consideration of several different performance measures. *Multi-objective* optimization is important when the use of objective functions associated with economics involves high uncertainty (e.g., in cases of thermodynamic analysis or prior to sizing) and is often replaced by the simultaneous consideration of objectives such as exergetic and energetic efficiency. This method is also important when sustainability objectives need to be considered simultaneously with economics. Sustainability considerations are often in conflict with economics because they increase the associated costs. Such trade-offs are unveiled using multi-objective optimization. Merits and shortcomings of such formulations as well as objective function types are summarized in Table 7.

In ORC design the choice of the objective function is very important. In some reviewed cases exergetic and thermal efficiency are used as design criteria but almost always together with some other index that is related with cost. From a thermodynamic perspective, two major and general objectives were mentioned in Section 4.3 as part of the work presented by Lecompte *et al.* [73] in terms of different ORC structures. In terms of economics, Novak Pintarič and Kravanja [77] mention that minimization of costs and maximization of profit are the most frequently used economic criteria in the design of industrial process systems. However, there are many other financial measures which can lead to different optimal solutions if applied in the objective function. Such measures involve the total annual cost (TAC), profit before taxes (PBT), payback time (PT), return on investment (ROI), net present value (NPV), internal rate of return (IRR) and equivalent annual cost (EAC). Novak Pintarič and Kravanja [78] extend their work to investigate the impact of using such criteria in *single-* and *multi-objective* optimization approaches. They break down the economic criteria into three classes: (a) Qualitative or non-monetary criteria (e.g., IRR or PT), (b) Quantitative or monetary criteria (e.g., Profit and TAC) and (c) Compromise criteria (e.g., NPV and TAC or Profit using depreciation with the annualization factor). The authors generally conclude that the NPV is the most appropriate objective function to use. Even so, sufficiently accurate process models are necessary so that the obtained results are both optimum and realistic.

- In *single-objective* optimization, they find that the Compromise criteria (NPV) are the most suitable because the obtained designs enable a fair compromise between profitability, operational efficiency, and environmental performance. The other criteria either favor solutions with small capital investment and cash flow but fast payback time and high profitability (Qualitative criteria) or vice versa (Quantitative criteria).
- In *multi-objective* optimization the NPV results in Pareto optimum designs that are close to the environmentally friendliest designs obtained by Quantitative criteria (e.g., Profit or TAC). On the other hand, the Qualitative criteria unveil environmental trade-offs in a much wider range.

5. Operation and Control of ORC Systems

5.1. Main Concepts and Computational Challenges

ORC systems operate in perpetually changing environments and therefore their operation should be constantly monitored and controlled. The main source of variation affecting the operation of ORC is the quality of the heat source. The heat source may experience changes in the flow rate and the temperature influencing the enthalpy content of the stream. Such changes would impact the degree of superheating in the outlet stream of the evaporator and the efficiency of the overall cycle. Other sources of variation in ORC are the efficiencies of the pump and the expander, and the heat transfer coefficients in the heat exchangers.

Feedback control is the main concept behind the maintenance of the controlled variables at predefined levels despite the influence of multiple and continuous disturbances. The key idea in feedback control is the utilization of the most recent information about the state of the plant through sensible and reliable measurements of the controlled variables. The controller actions are determined using the calculated deviation of the controlled variables from predefined set points (*i.e.*, reference points). The main objective of process control remains the transfer of process variability from the most important in terms of profitability and product quality process streams and variables to process streams and variables of reduced importance. Such streams that are the recipients of the variability on valuable and therefore important variables are usually utility and auxiliary streams (e.g., air or water cooling streams, bypass streams, working fluid flowrate). For instance, in an ORC system variability in the heat source is transferred to the electric power generator which is attached to the expander. Electric power is usually intended to satisfy a critical specification on the power load. Therefore, variability in the power generation may be attenuated by the control system by manipulating the flow rate of the working fluid and/or the expander bypass stream. Obviously, either action would also affect the working fluid condensation and the cooling requirements in the condenser usually imposed by an air cooling system.

Feedback control operates in order to correct any deviations of the controlled variables from predefined set point levels after the effects of exogenous disturbances on the controlled variables has been sensed by the measurement sensors. The controller action is computed based on the calculated deviation from a pre-defined set point (i.e., error in the controlled variables). Linear analysis of the outlined dynamic system with either Laplace transform or state space formulation are the most commonly used practices to analyze and investigate the process dynamics and interactions [79]. Overall plant dynamics include the dynamics of the associated process units such as heat exchangers, pumps, expanders and so forth, the implemented controllers, the incorporated actuators, and the installed sensors. The process representation by transfer functions through a Laplace transform of the governing differential equations enables the evaluation of the system dynamic characteristics. Alternatively, a state space representation enables the representation of multiple input, multiple output systems. Several controller design methods are available that aim to achieve the desired dynamic performance for the system [80].

Real time control applications are usually based on a control law that has been offline calculated. In this aspect, online calculations are limited to the evaluation of the control actions in a multi-loop fashion, where one manipulated variables is used to regulate one controlled variable, with minimal computational effort. This feature enables the implementation of a relatively small control interval; the time interval that a new control action is calculated and implemented in the system. However, such control systems must be designed with provisions to perform adequately even though the process has shifted away from the nominal operating point (e.g., due to a change in the power level) or process parameters have varied significantly during operation (e.g., due to fouling in the heat exchanger or other process equipment malfunctions). On the contrary, model-based control systems utilize at real time process model predictions that enable the controller to allocate the control effort in multi-variable systems optimally [81]. The achieved controller dynamic performance can be significantly improved over multi-loop approaches because input-output interactions are explicitly taken into consideration but at the expense of increased computational effort. Basically, dedicated control system can easily manage the involved computational effort, especially when the employed models are linear [82]. Nonlinear model predictive control systems [83] offer definitely improved accuracy of model predictions and therefore better control performance but require specialized solution algorithms for optimization and state estimation [84]. A schematic of the relationship between process detail involved in online control applications with the associated control effort is provided in Figure 6.

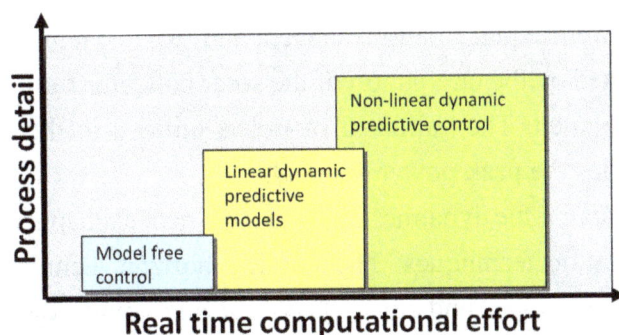

Figure 6. Control approaches with respect to process detail and real time computational effort.

5.2. Dynamic Models

Identifying the dynamics of ORC systems is important in the design and achieved performance of the control system. Process models that are based on first principles arising from the physical phenomena (e.g., heat transfer, compression, expansion) taking place in the ORC provide the most reliable and accurate description of the system behavior. The models are basically consisted of material, energy and momentum balances in dynamic mode accompanied with constitutive equations. However, the models involve a number of parameters associated with the physical and chemical phenomena (e.g., heat and mass transfer coefficients, expander and pump efficiencies, physical properties and so forth). The estimation of the model parameters requires the collection of experimental data from well-designed experiments with sufficiently rich information in calculating accurately the model parameters. The most reliable way for the parameter estimation is the fitting of the model response to the dynamic data using maximum likelihood principles and dynamic programming techniques. However, the development of a detailed mechanistic model can be replaced by empirical modeling performed using input-output data. This simplified technique requires the execution of experimental step changes in the input process variables while maintaining all other variables in manual operation [85]. The magnitude of the step change depends on the process nonlinearity and the measurement noise level in the measured variables. Depending on the shape of the output response of the process to an input step change, the order of the dynamic system can be identified. Most dynamic systems can be approximated as first-order models with dead-time [79]. Dead-time is the time it takes to observe the effect of an input signal change in the output variables. High order over-damped systems resemble the behavior of a first-order plus dead-time model and therefore it becomes an attractive modeling option. The estimation of the model parameters for such a model; namely the process gain, the time constant, and the dead-time, can be easily performed [86]. An alternative empirical model building is based on time series analysis [87]. Auto- and cross-correlation of time series can be utilized for the identification of the process model order whereas ordinary least squares and recursive least squares can be used for the estimation of the model parameters. Zhang et al. [88–90] have employed auto-regressive integrated moving average models in the control of ORC systems.

In a typical ORC system the main source of dynamic characteristics are the evaporator and the condenser. The evaporator is a heat exchanger with single phase (preheating and superheating) and double phase (evaporation) regions. Twomey et al. [91] developed a dynamic model for a solar ORC using a scroll expander. The dynamic model based on first principles showed good agreement with experimental data regarding power output, rotational speed, and exhaust temperature. The effect of the tank volume which is being heated by the circulation of the solar collector fluid and is acting as the heat source for the system is investigated. The validated model is utilized in the design of solar thermal cogeneration systems that satisfies the peak power demand.

Wei *et al.* [92] attempted to capture the dynamics of the system heat exchangers using models based on moving boundary and discretization techniques. The moving horizon technique aims to identify the boundaries between a single (liquid or gas) and a two phase (gas and liquid) region within the evaporator by imposing explicit energy balances. In the discretization technique a number of computational cells is introduced within each region with the appropriate boundary conditions. The two methods are compared in terms of accuracy, complexity and simulation speed with the moving boundary technique exhibiting better characteristics for online control applications. However, the discretization methods appeared to be more suitable for the simulation of start-up and shut-down conditions.

Similarly, Bamgbopa and Uzgoren [93] developed a dynamic model for the heat exchangers and static models for the pump and the expander and studied the power output for varying flow rate and temperature for the hot and cold sources in a system that employed R245fa as the working fluid. In a subsequent article by Bamgbopa and Uzgoren [94] the models were utilized to evaluate the steady state efficiency of a solar ORC system. The values for the decision variables (hot source flow rate and temperature, and working fluid flow rate) that maximize the overall efficiency of the system were determined. Regression models were developed to characterize the effectiveness of the system in terms of the ratio of the working fluid flowrate to the heat source flow rate and the heat source temperature at the inlet of the evaporator. Table 8 summarizes the employed modeling approaches and the purpose of the developed model.

Table 8. Dynamic modeling approaches.

Authors	Modeling Method	Equipment	Purpose
Quoilin *et al.* [85]	Empirical (regression)	Entire ORC system	Control system design
Zhang *et al.* [88–90]	Empirical (regression)	Entire ORC system	Control system design
Wei *et al.* [92]	First principles	Entire ORC system	Start-up and shut-down simulations
Bamgbopa and Uzgoren [93]	First principles	Heat exchangers (dynamic), pump expander (static)	Power output computation
Bamgbopa and Uzgoren [94]	First principles	Solar ORC	Steady state efficiency

5.3. Control Approaches

Control systems for ORC can be generally categorized in multi-loop and multi-variable schemes. Quoilin *et al.* [95] proposed a series of control strategies for an ORC. Initially, a static model was used to determine the optimal evaporating temperature and superheating for a wide range of heat source and heat sink conditions. The manipulated variables in the system included the expander speed and the pump capacity. The optimal evaporator temperature was derived from a regression model and was followed by the control system consisted of two proportional-integral controllers. In another version of the control scheme a correlation was utilized for the pump capacity based on expander speed of rotation and the heat and cool source temperature. This acts as a feedforward control system as the measurement of the expander speed was used to provide the set point for the working fluid flow rate. In this way the response of the control system was significantly faster. However, proper tuning of the controllers should be maintained in order to achieve stability. Simulated results verified that the control scheme that follows the optimal evaporation temperature trajectory exhibited superior performance. Peralez *et al.* [96] used a model based control scheme considering system inversion for the control of the superheating temperature which affects both cycle performance and system safety. The model inversion introduces a feedforward action in addition to the feedback controller to effectively compensate for disturbances in the evaporator. Kosmadakis *et al.* [97] discussed potential control strategies for double stage expanders in ORC systems.

Multivariable control schemes have attracted the attention of researchers in the control of ORC systems because of the superior performance they exhibit in ORC applications [98]. Zhang *et al.* [88] developed a dynamic model with moving boundaries for the evaporator and the condenser. Subsequently, a linear state space model was derived for control system design purposes. The control objectives were the minimization of system interaction in order to achieve good disturbance rejection and the maximization of the overall system efficiency. For the latter, the degree of superheating in the evaporator and the condenser outlet temperatures were regulated. A linear quadratic regulator coupled with a PI (proportional-integral) controller have been designed and simulated for set point changes in the power output and the throttle valve pressure (*i.e.*, pressure at the entrance of the expander) as well as the superheating and condenser temperatures. The PI controller maintained the condenser outlet temperature at the desired level. Similarly, disturbance rejection scenarios were investigated associated with hot gas stream velocity variation and throttle valve dynamics. Zhang *et al.* [99] extended the previous work by developing an extended observer that aims to provide accurate state estimates for the system.

Zhang *et al.* [89] introduced a dynamic model for a waste heat recovery system based on ORC with R245fa as the working fluid. The first principles dynamic model was then converted to a CARIMA (controlled auto-regressive integrated moving average) model for use in a model predictive control scheme. The controlled variables in the multi variable control scheme were the system power output, the evaporator pressure, the superheating temperature, and the condenser temperature. These variables were controlled using the pump and expander rotating speeds and the air flow in the condenser. A constrained generalized predictive controller [100] was implemented which rejected disturbances and followed set point effectively. In a subsequent paper, Zhang *et al.* [90] introduced a constrained generalized predictive controller that considered bounds on both the manipulated and controlled variables as well as the rate of change for the manipulated variables. The performance of the controller has been evaluated for disturbances in the temperature and the flow rate of the heat source stream. Power output was maintained at the desired level despite the disturbances. Additionally, set point changes for the evaporator pressure, the superheating temperature and the condenser temperature were successfully tracked by the controller.

In a recent work Hou *et al.* [101] introduced a minimum variance controller with real-time parameter estimation for a CARMA (controlled auto-regressive moving average) model. A recursive least squares technique was implemented for the parameter estimation. However, the proposed control scheme does not consider a model for the stochastic disturbances in the system. Uncertainties may play a significant role in the performance of the ORC system. Therefore, changes in the dynamic features should be monitored on-line using the measurements from the process. Additionally, the inherent nonlinearities in the system may make the predictions from linearized process models highly inaccurate. To this end,

Zhang *et al.* [88] proposed a state extended observer for the on-line update of states and model parameters. The updated model is then utilized in a linear quadratic regulator with a PI controller for the plant control.

Table 9 summarizes the literature in control strategies of ORC systems. In conclusion, multi-loop control systems works efficiently when good conceptual and process knowledge is utilized based on prior system analysis. Multi-loop control systems are easily implemented and maintained but may require frequent tune-up to account for process changes and operating condition variations. Multi-variable control systems require the development of a dynamic process model that can provide accurate process behavior predictions over a wide range of operation conditions. The implementation is definitely more challenging but guarantees good control performance through the explicit consideration of process interactions.

Table 9. Control approaches.

Authors	System Type	Control Approach	Manipulated/Controlled Parameters
Quoilin *et al.* [95]	Low grade ORC waste heat recovery	PID (multi-loop)	Pump speed, expander speed/evaporating temperature, superheating
Peralez *et al.* [96]	ORC waste heat recovery	Nonlinear model inversion	Exhaust gas by-pass valve, expander by-pass valve, pump speed, expander speed/Superheating temperature
Hou *et al.* [101]	ORC waste heat recovery	Minimum variance controller (multi-variable)	Throttle valve position, mass flow rate of working fluid, mass flow rate of exhaust gas, air flow rate/power, throttle pressure, evaporator outlet temperature, condenser outlet temperature
Zhang *et al.* [88,99]	ORC waste heat recovery	Linear Quadratic Regulator with extended observer (multi-variable)	Throttle valve position, working fluid pump speed, exhaust gas velocity, air velocity/power, throttle pressure, evaporator outlet temperature, condenser outlet temperature
Zhang *et al.* [89]	ORC waste heat recovery	Model predictive control (multi-variable)	Throttle valve position, working fluid pump speed, exhaust gas velocity, air velocity/power, throttle pressure, evaporator outlet temperature, condenser outlet temperature
Zhang *et al.* [90]	ORC waste heat recovery	Constrained generalized predictive controller (multi-variable)	Throttle valve position, working fluid pump speed, exhaust gas velocity, air velocity/power, throttle pressure, evaporator outlet temperature, condenser outlet temperature, Constraints on system variables.

Part load operation away of the nominal design ORC settings is an important issue that requires the utilization of efficient control methods. The operation of the ORC system at part load conditions also requires the modeling of the thermal efficiency with respect to the off-design operating conditions. We review some part load system analyses here however all works do not consider feedback control system performance but rather focus on steady state operation.

Ibarra *et al.* [102] focused on the characterization of the expander and the heat recuperator to obtain an accurate representation of the optimal part load conditions. The study involved a number of different working fluids. Manente *et al.* [103] provided correction factors for the turbine isentropic efficiency due to variations of the isentropic enthalpy drop and the working fluid mass flow rate from the design point. In this way the calculation of the sensitivity of performance indicators with respect to the off-design point was possible. The control system utilized subsequently the optimal operation point at part load. A cascade type of control has been implemented in the combined gas turbine—ORC system by de Escalona *et al.* [104]. The study focused on the benefits from the addition of the ORC for waste heat recovery and considered part load conditions in the ORC performance. Additional latest works and applications are reviewed in a recent work addressing the part load performance of a wet indirectly fired gas turbine integrated with an ORC turbogenerator [105].

5.4. Remarks on Employed Methods

Control of ORC systems enables the efficient compensation of the effect disturbances have on the power output and guarantee the equipment operation within safety limits.

- Multi-loop control systems are relatively simple to implement but require careful tuning to enable stable and acceptable dynamic performance. Highly interactive systems hinder the achievable control performance and therefore the introduction of model based control techniques becomes a viable option. System interaction is further increased whith more complex ORC configurations (e.g., multi-pressure or multi-temperature systems, multiple expansion units and so forth).
- Model based control systems require the development of accurate dynamic models for the individual subsystems. Usually, linear models with suitable disturbance models and integral action can meet the control objectives. The model development effort is accompanied by the execution of well-designed experiments in order to estimate model parameters and validate the model structure and predictions. In addition, an online parameter estimation procedure is attached to the feedback loop so that the control models can adapt to plant drifts. Model based control usually results in improved dynamic performance as process interaction is taken into consideration explicitly but model accuracy is an essential factor for acceptable set-point tracking and disturbance rejection.

6. ORC Integration with Multiple Heat Source Streams

6.1. Main Concepts

The work reviewed so far has focused on working fluid selection, ORC design optimization and control with respect to a single heat source and sink. Energy intensive industrial processes often require significant amounts of low to medium grade heat to be removed into cooling water or another cooling medium, which could be utilized through synergies with surrounding processes and sectors [106]. ORCs offer a potentially promising route to monetize this waste heat through conversion to power. Besides the widespread industrial processes, other applications with multiple heat sources have been identified. Romeo *et al.* [107] integrate the multiple intercoolers in compression trains with ORC configurations. The work designs the cycles (high and low pressure) together with the compression train to match intercooling waste heats and the ORCs. The work demonstrates significant energy savings of over 10% from the integration on the ORCs with the compression train. In another application, Soffiato *et al.* [108] integrate ORCs with the available waste heat streams onboard a LNG carrier. The work shows that power output can be increased by 3.5% through ORC integration. DiGenova *et al.* [2] study the integration of ORCs with an energy intensive Fischer-Tropsch plant to convert coal to liquid fuels. They apply the Pinch Analysis techniques to explore the performance of single and multi-pressure cycles to convert heat from process streams to power and observe that the carefully integrated ORCs significantly outperform steam cycles in terms of conversion efficiency.

Although systematic approaches for the integration of ORC systems with multiple heat source streams are only emerging, the general field of process energy integration, in which most proposed approaches have their roots, is well established. In the 1970s, energy integration approaches emerged with the advent of Pinch Analysis for targeting minimum process heat requirements and heat recovery network design [109]. The methods are well established and routinely applied in the design of chemical processes [110], which has led to significant energy savings in the process industries. These approaches have their origin in thermodynamic analysis and provide graphical representations of the design problem to guide the analysis of energy flows and gain insights into promising heat recovery and power generation options [111]. To enable the better screening of design options and to incorporate economic criteria in decision making, numerous complementary approaches based on mathematical optimization have emerged over time to explore both operational design decisions as well as structural design alternatives for process heat and power systems. Smith [110] provides an overview of established energy integration approaches. The graphical approaches do not present computational challenges, whereas the optimization-based process integration approaches incorporate similar challenges to those reported in Section 4.

6.2. Reviewed Approaches

The integration of ORCs with multiple heat source streams and in the context of process heat and power generation systems has had no reference in the literature until very recently. Over the past five years, systematic approaches to guide the efficient integration of ORC systems have started to emerge (Tables 10 and 11). Hackl and Harvey [112] employ Total Site Analysis for power production from low temperature excess process heat from a chemical cluster using a simple ORC. Desai and Bandyopadhyay [113] were the first to study the integration of ORCs with a background process of multiple potential heat source streams. They adopt graphical approaches and apply established Pinch analysis techniques in the form of Grand Composite Curves to explore ORC integration targets and develop heat exchanger network designs to achieve them. The work highlights the strong dependence of high performance ORC integration strategies on the specific characteristics of the background process. With a focus on site utility systems with multiple steam levels and turbines, Kapil *et al.* [114] introduce a co-generation targeting method that considers the optimization of pressure levels together with integration options for ORCs and heat pumps as low grade heat utilization options.

Hipolito-Valencia *et al.* [115] propose a superstructure approach to capture various possible heat transfer options between process streams and the ORC. Similar to the work by Desai and Bandyopadhyay [113], the approach focusses on the efficient integration of the ORC with the multiple source streams of the background process. In a subsequent contribution, Hipolito-Valencia [116] propose an approach for interplant energy integration that considers ORCs for power generation. Lira-Barragán et al. [117] continue this approach to select the best possible conditions, heat exchanger network configuration and type of process in a trigeneration system. The process types involve a steam Rankine cycle, an ORC and an absorption refrigeration system. The authors consider economic, environmental and social indices as objective functions in optimization. All three works resulted in MINLP formulations that were searched using deterministic optimization solvers. Neither approach optimizes the expansion section of the ORC.

Gutiérrez-Arriaga et al. [118] proposed an approach for energy integration involving waste heat recovery through an ORC which is based on a two-stage procedure. In the first stage, heating and cooling targets are determined through heat integration. This enables the identification of the excess process heat available for use in the ORC. The optimization of the operating conditions and design of the cogeneration system are carried out in the second stage using Genetic Algorithms.

Table 10. Optimization-based heat source integration approaches.

Authors	Integration Approach	Evaluated Options	Integration/Design Criteria
Kwak et al. [119]	Total site analysis, optimization of ORC operation	Sixteen working fluids, turbine inlet temperature condenser outlet temperature	Total annualized cost
Chen et al. [120]	Superstructure-based optimization of HEN integrated with ORC, MINLP solver	Number and connections of HEN, operating parameters of HEN and ORC	Generated ORC work
Marechal and Kalitventzeff [121]	Mathematical model of exergy composite curves (MILP solver), ORC operation optimization and fluid selection (MILP solver)	List of few pre-selected fluids, utility flowrates, several ORC operating characteristics	Costs, exergy losses
Soffiato et al. [108]	SQP solver (deterministic) for ORC optimization in an iterative procedure, pinch composite curves for ORC-heat source matching, evaluation of each ORC structure and working fluid combination	Six working fluids, three pre-selected structures (simple cycle, regenerative cycle, and two-stage cycle), the evaporation pressures and the degrees of superheating in one or two stages, the ratio between the mass flow rates in the two stages	Net ORC power output
Lira-Barragán et al. [117]	Multi-objective MINLP, results reported for all working fluids	Three working fluids, structure and operating characteristics of heat exchanger network, existence of ORC and/or absorption refrigeration system	Economic (annual profit), environmental (greenhouse gas emissions), social (number of jobs generated)
Gutiérrez-Arriaga et al. [118]	Pinch grand composite curves, Genetic Algorithms to optimize operation of a basic ORC, results reported for 3 different working fluids	Three working fluids, operating ORC parameters	Gross annual profit
Kapil et al. [114]	Total site analysis (NLP optimization), ORC process simulation	Pressure of different steam levels	Enthalpy difference of shifted heat sink and source, thermal efficiency, purchase cost
Hipólito-Valencia [115,116]	Heat exchanger network superstructure, MINLP solver	Total heat transfer area, network configuration, operating parameters, two working fluids	Total annualized cost

Kwak *et al.* [119] investigate different technologies, including ORCs for energy recovery and exploitation in different industrial sites. The authors perform a Total Site Analysis to identify energy recovery targets and then identify the optimum ORC operating parameters together with the working fluid (from a list of 16 pre-defined fluids) in order to best recover the available energy.

Chen *et al.* [120] present a mathematical model for the synthesis of a heat-exchanger network (HEN) which is integrated with an organic ORC for the recovery of low-grade industrial waste heat. An ORC-incorporated stage wise superstructure considering all possible heat-exchange matches between process hot/cold streams and the ORC is first presented. First, a stand-alone HEN is synthesized to minimize the external utility consumption. An ORC is then incorporated into the HEN with the objective of maximizing the work produced from waste heat (without increasing the use of a hot utility. The problem is formulated and solved as a two stage MINLP.

Marechal and Kalitventzeff [121] proposed a method for the investigation of ORC process characteristics which is based on the analysis of the shape of the grand composite curve, combined with the use of the minimum exergy losses concept, heuristic rules and a cost optimisation technique. First, the recovery targets of the background (waste-heat) industrial process are determined through an MILP-based optimization to minimize exergy losses using the utility flowrates as decision variables. ORCs are then designed together with working fluids selected from a pre-specified list to optimally match the identified energy recovery opportunities. The identified ORCs are characterized in terms of the condenser and evaporator temperatures and pressure conditions, the opportunity for superheating, the expected flow-rate and efficiency of the cycle. The non-linear cost estimation of the condensers, boilers, turbines and pumps are linearized and the best matches of the designed ORCs with the background process are identified using MILP-based optimization to minimize costs. The focus of the proposed developments is on the integration of the ORC vaporization and condensation sections.

Stijepovic *et al.* [70] adopt the exergy composite curves (ECCs) approach developed by Linnhoff and Dhole [71] to explore the potential for ORC process improvements through better utilization of the available heat. The ECC shape reflects on ORC operating conditions which may be interpreted by different process configurations (e.g., simultaneous consideration of different pressure levels may require multiple turbines interconnected at various heat exchanger topologies to match the necessary temperatures). Details of this approach have been reviewed in Table 6.

Most recently, Song *et al.* [122] explores integration schemes for single and dual ORCs with multiple waste heat streams through simulation. The work identifies the dual cycle as the best performing configuration for a refinery case study. This highlights the need to develop optimal ORC integration methods in the future that can take into account multiple heat source streams and multiple integrated power cycles simultaneously.

Tchanche *et al.* [123] developed an approach to evaluate the performance of different ORC configurations by using several criteria based on exergies for different parts of the equipment. Using graph theory they conceptualized the exergy flows and losses within different sections of an ORC system, investigating three different cycle topologies in the condensing and pumping sections (*i.e.*, regenerative heat exchanger, open feed liquid heater and closed feed liquid heater).

Yu *et al.* [124] propose a new method to simultaneously determine the working fluid and operating conditions in an ORC. The Preheating Pinch Point and the Vaporization Pinch Point are introduced. The method is based on a newly defined parameter named "predictor" that can predict the pinch position between the waste heat carrier and the working fluid, calculate the mass flow rate of working fluid and the amount of heat recovered, and determine the optimum working fluid and corresponding operating conditions simultaneously. The authors consider 11 pre-selected working fluids which are considered as decision options simultaneously with the process features. The objective is to maximize the power output without considering the equipment cost and operating expenses.

Safarian and Aramoun [125] employ a combined energy- and exergy-based analysis approach to evaluate four ORC configurations, namely a basic ORC, an ORC with turbine bleeding, with regeneration and with both turbine bleeding and regeneration. The authors employ several analysis criteria, calculate exergy losses and find that the evaporator has major contribution in the exergy destruction which is improved by increase in its pressure. Furthermore, the configuration with turbine bleeding and regeneration enables a maximization of thermal and exergetic efficiencies and minimization of exergy losses.

Luo et al. [126] present s systematic hybrid methodology of graphical targeting and mathematical modeling to address the optimum integration of a regenerative ORC in a steam network. The objective function is to minimize the fuel consumption of the steam power plant. The terminal temperature and heat load of the process-heated boiler feed water are the two decision variables. The graphical targeting method is proposed to ascertain the bounds and constraints of the two decision variables. A mathematical model incorporating rigorous simulations of the turbine is formulated to achieve the optimal heat integration scheme.

Table 11. Graphical or simulation-based heat source integration approaches.

Authors	Integration Approach	Evaluated Options	Integration/Design Criteria
Yu et al. [124]	Pinch-based energy recovery targeting, iterative enumeration	Eleven working fluids, cycle operating parameters	Power output
Safarian and Aramoun [125]	Exergy- and energy-based analysis to identify best ORC structure, evaluation of each structure separately	Basic ORC, ORC incorporating turbine bleeding, regenerative ORC, ORC incorporating both turbine bleeding and regeneration	Degree of thermodynamic perfection, exergetic efficiency, thermal efficiency, influence coefficient, exergy loss
Luo et al. [126]	Pinch-based energy targeting, simulation-based objective function evaluation	Terminal temperature and heat load of the process-heated boiler feed water	Fuel consumption of steam power plant
Tchanche et al. [123]	Energy and exergy flow analysis, evaluation of each working fluid and ORC structure combination separately through simulations	Four working fluids, four structures (basic ORC, ORC with regenerative heat exchanger, with open or closed feed liquid heater)	Exergy losses, degree of thermodynamic perfection, exergetic efficiency, energetic efficiency
Romeo et al. [107]	Pinch-based energy integration, different temperature levels, evaluation of ORC structure for each working fluid	Pre-selected two-stage (dual pressure) ORC, temperature of waste heat source, six pre-selected working fluids	Energetic efficiency
DiGenova et al. [2]	Pinch composite curves for ORC-heat source matching options, evaluation of each structure separately	Five ORC structures (basic ORC, with reheat stages, with multiple pressure levels, recuperator and balanced recuperator)	Thermal efficiency

Table 11. *Cont.*

Authors	Integration Approach	Evaluated Options	Integration/Design Criteria
Hackl and Harvey [112]	Total site analysis, ORC simulations with different working fluids	Five pure fluids, one mixture, ORC operating parameters	Net excess heat, cost of electricity, payback period of investment, CO_2 emissions reduction
Desai and Bandyopadhyay [113]	Pinch-based graphical integration, ORC simulations with different fluids	Sixteen fluids, basic ORC, ORC with turbine bleeding and regeneration, heat exchanger network configuration and conditions	Net work output, thermal efficiency
Song *et al.* [122]	Matching of heat sources at different temperatures with different ORC configurations and fluids, simulations	Eight pure and six mixed working fluids, Dual integrated or independent ORC, single ORC	Net power output, heat transfer area and their ratio

6.3. Remarks on Integration

Despite the recent progress, there remains significant scope to develop more widely applicable, systematic approaches to guide the optimal integration of ORCs with multiple heat streams in the context of background processes. Future developments are expected to focus on a number of unexplored aspects of the ORC integration problem:

- Enriched representations of ORC configurations and multiple heat steams in the form of superstructures to provide a representation of all possible alternative configurations, including multi-pressure and multi-cycle systems.
- Multi-scale approaches to bridge the gap between higher-level ORC integration and detailed design decisions impacting on performance such as heat exchanger design optimization.
- Multi-scale approaches to support integrated decision making across the working fluid selection, ORC design and ORC integration problem levels, and
- Global search schemes for ORC integration with multiple heat sources similar to the approaches that emerge for ORC design optimization.

An important observation is that most existing works consider energy analysis, although the combination of energy and exergy analysis is also very useful for the investigation of different ORC configurations and integration options. Usually, energy analysis takes into account only wasted heat which is available at temperatures where it may be re-utilized as heat (e.g., steam). However, wasted heat of low enthalpy content may be transformed into power through ORC and re-utilized in the process. Exergy analysis is particularly relevant in these cases as it focuses on the maximum useful work that can be produced from a heat source. When heat is transferred part of the thermal energy is degraded due to process irreversibility. The key is to minimize the part of thermal energy lost due to degradation. Exergy is the maximum quantity of work that can be produced in a cyclic thermodynamic process. It captures only that part of the thermal energy which may be transformed into work. Energetic analysis targets the recovery of heat loads. Exergetic analysis targets the maximum work that may be recovered from an available heat source, hence it may target exergy loss or the ideal work equivalent lost in heat transfer.

7. Concluding Remarks

The design of ORC systems is a challenging task. Many design alternatives exist at each level of design, from working fluid selection via cycle optimization and control through to the efficient integration of the cycle with background processes. A number of systematic approaches have emerged over the past few years that aim at supporting the designer in making optimal choices at each level of ORC development. This paper has aimed to provide a state-of-the-art overview of these emerging approaches with a particular emphasis on computer-aided design methods and aimed to highlight areas that may benefit from further research and development.

Author Contributions

The three authors jointly developed this review article. Individual focus areas have been computer-aided ORC working fluid design and selection (A.I. Papadopoulos), ORC design and integration (P.Linke) and ORC operation and control (P. Seferlis).

Nomenclature

CAMD	Computer-Aided Molecular Design
CARMA	Controlled auto-regressive moving average
CFD	Computational fluid dynamic
CoMT	Continuous molecular targeting
COSMO-RS	Conductor-like screening model for real solvents
DFT	Density functional theory
EAC	Equivalent annual cost
EoS	Equation of state
GC	Group contribution
GWP	Global warming potential
HEN	Heat exchanger network
IRR	Internal rate of return
MILP	Mixed integer linear programming
MNLP	Mixed integer non-linear programming
NLP	Non-linear programming
NPV	Net present value
ODP	Ozone depletion potential
ORC	Organic Rankine cycle
PBT	Profit before taxes
PC-SAFT	Perturbed chain statistical associating fluid theory
PI	Proportional-integral
PID	Proportional-integral-derivative
PT	Payback time
QSPR	Quantitative structure-property relationships
ROI	Return on investment
SQP	Sequential quadratic programming
TAC	Total annual cost

Conflicts of Interest

The authors declare no conflict of interest.

References

1. Tchanche, B.F.; Lambrinos, G.; Frangoudakis, A.; Papadakis, G. Low-grade heat conversion into power using organic Rankine cycles—A review of various applications. *Renew. Sustain. Energy Rev.* **2011**, *15*, 3963–3979.

2. DiGenova, K.J.; Botros, B.B.; Brisson, J.G. Method for customizing an organic Rankine cycle to a complex heat source for efficient energy conversion, demonstrated on a Fischer Tropsch plant. *Appl. Energy* **2013**, *102*, 746–754.

3. Hung, T.C. Waste heat recovery of organic Rankine cycle using dry fluids. *Energy Convers. Manag.* **2001**, *42*, 539–553.

4. Papadopoulos, A.I.; Stijepovic, M.; Linke, P. On the systematic design and selection of optimal working fluids for organic Rankine cycles. *Appl. Therm. Eng.* **2010**, *30*, 760–769.

5. Papadopoulos, A.I.; Stijepovic, M.; Linke, P.; Seferlis, P.; Voutetakis, S. Power generation from low enthalpy geothermal fields by design and selection of efficient working fluids for organic Rankine cycles. *Chem. Eng. Trans.* **2010**, *21*, 61–66.

6. Preißinger, M.; Heberle, F.; Brüggemann, D. Advanced organic Rankine cycle for geothermal application. *Int. J. Low Carbon Technol.* **2013**, doi:10.1093/ijlct/ctt021.

7. Mavrou, P.; Papadopoulos, A.I.; Stijepovic, M.; Seferlis, P.; Linke, P.; Voutetakis, S. Novel and conventional working fluid mixtures for solar Rankine cycles: Performance assessment and multi-criteria selection. *Appl. Therm. Eng.* **2015**, *75*, 384–396.

8. Schuster, A.; Karellas, S.; Kakaras, E.; Spliethoff, E. Energetic and economic investigation of innovative organic Rankine cycle applications. *Appl. Therm. Eng.* **2008**, *29*, 1809–1817.

9. Tsoka, C.; Johns, W.R.; Linke, P.; Kokossis, A. Towards sustainability and green chemical engineering: Tools and technology requirements. *Green Chem.* **2004**, *8*, 401–406.

10. Biegler, L.T.; Grossmann, I.E.; Westerberg, A.W. *Systematic Methods of Chemical Process Design*; Prentice Hall International Series in the Physical and Chemical Engineering Sciences; Prentice Hall PTR: Upper Saddle River, NJ, USA, 1997.

11. Klemeš, J.J. *Process Integration Handbook*; Woodhead Publishing: Cambridge, UK, 2013.

12. Seferlis, P.; Georgiadis, M.C. *The Integration of Process Design and Control*; Computer Aided Chemical Engineering, 17; Elsevier Science B.V.: Amsterdam, The Netherlands, 2004.

13. Adjiman, C.S. Optimal solvent design approaches. In *Encyclopedia of Optimization*, 2nd ed.; Floudas, C.A., Pardalos, P.M., Eds.; Springer: New York, NY, USA, 2009; pp. 2750–2757.

14. Adjiman, C.S.; Galindo, A. Molecular systems engineering. In *Process Systems Engineering*; Pistikopoulos, E.N., Georgiadis, M.C., Dua, V., Eds.; Wiley: Weinheim, Germany, 2010; Volume 6.

15. Ng, L.Y.; Chong, F.K.; Chemmangattuvalappil, N.G. Challenges and opportunities in computer-aided molecular design. *Comput. Chem. Eng.* **2015**, doi:10.1016/j.compchemeng.2015.03.009.

16. Papadopoulos, A.I.; Linke, P. Integrated solvent and process selection for separation and reactive separation systems. *Chem. Eng. Process. Process Intensif.* **2009**, *48*, 1047–1060.

17. Stijepovic, M.; Linke, P.; Papadopoulos, A.I.; Grujic, A. On the role of working fluid properties in organic Rankine cycle performance. *Appl. Therm. Eng.* **2012**, *36*, 406–413.

18. Papadopoulos, A.I.; Stijepovic, M.; Linke, P.; Seferlis, P.; Voutetakis, S. Toward optimum working fluid mixtures for organic Rankine cycles using molecular design and sensitivity analysis. *Ind. Eng. Chem. Res.* **2013**, *52*, 12116–12133.

19. Papadopoulos, A.I.; Stijepovic, M.; Linke, P.; Seferlis, P.; Voutetakis, S. Molecular design of working fluid mixtures for organic Rankine cycles. *Comput. Aided Chem. Eng.* **2013**, *32*, 289–294.

20. Bao, J.; Zhao, L. A review of working fluid and expander selections for organic Rankine cycle. *Renew. Sustain. Energy Rev.* **2013**, *24*, 325–342.

21. Angelino, G.; di Paliano, P.C. Multicomponent working fluids for organic Rankine cycles (ORCs). *Energy* **1998**, *23*, 449–463.

22. Demuth, O.J.; Kochan, R.J. *Analyses of Mixed Hydrocarbon Binary Thermodynamic Cycles for Moderate Temperature Geothermal Resources Using Regeneration Techniques*; Technical Report; Idaho National Engineering Lab.: Idaho Falls, ID, USA, 1981. Available online: http://www.osti.gov/geothermal/servlets/purl/5281969-D6H9jj/native/5281969.pdf (accessed on 7 April 2015).

23. Gawlik, K.; Hassani, V. Advanced binary cycles: Optimum working fluids. In Proceedings of the 32nd Intersociety Energy Conversion Engineering Conference, Honolulu, HI, USA, 27 July–1 August 1997; Volume 3, pp. 1809–1814.

24. Borsukiewicz-Gozdur, A.; Nowak, W. Comparative analysis of natural and synthetic refrigerants in application to low temperature Clausius-Rankine cycle. *Energy* **2007**, *32*, 344–352.

25. Angelino, G.; di Paliano, P.C. Air cooled siloxane bottoming cycle for molten carbonate fuel cells. In Proceedings of the Fuel Cell Seminar, Portland, OR, USA, 30 October–2 November 2010.

26. Bliem, C. Zeotropic mixtures of halocarbons as working fluids in binary geothermal power generation cycles. In Proceedings of the 22nd Intersociety Energy Conversion Engineering Conference, Portland, OR, USA, 10–14 August 1987. Available online: http://www.osti.gov/bridge/servlets/purl/5914218-ULxh0x/5914218.pdf (accessed on 7 April 2015).

27. Wang, X.D.; Zhao, L. Analysis of zeotropic mixtures used in low-temperature solar Rankine cycles for power generation. *Sol. Energy* **2009**, *83*, 605–613.

28. Heberle, F.; Preißinger, M.; Brüggemann, D. Zeotropic mixtures as working fluids in organic Rankine cycles for low-enthalpy geothermal resources. *Renew. Energy* **2012**, *37*, 364–370.

29. Chys, M.; van den Broek, M.; Vanslambrouck, B.; de Paepe, M. Potential of zeotropic mixtures as working fluids in organic Rankine cycles. *Energy* **2012**, *44*, 623–632.

30. Victor, R.A.; Kim, J.-K.; Smith, R. Composition optimisation of working fluids for organic Rankine cycles and Kalina cycles. *Energy* **2013**, *55*, 114–126.

31. Woodland, B.J.; Krishna, A.; Groll, E.A.; Braun, J.E.; Travis Horton, W.; Garimella, S.V. Thermodynamic comparison of organic Rankine cycles employing liquid-flooded expansion or a solution circuit. *Appl. Therm. Eng.* **2013**, *61*, 859–865.

32. Elsayed, A.; Embaye, M.; Al-Dadah, R.; Mahmoud, S.; Rezk, A. Thermodynamic performance of Kalina cycle system 11 (KCS11): Feasibility of using alternative zeotropic mixtures. *Int. J. Low Carbon Technol.* **2013**, *8* (Suppl. S1), i69–i78.

33. Micheli, D.; Pinamonti, P.; Reini, M.; Taccani, R. Performance analysis and working fluid optimization of a cogenerative organic rankine cycle plant. *J. Energy Resour. Technol.* **2013**, *135*, 021601.

34. Ramachandran, K.I.; Deepa, G.; Namboori, K. *Computational Chemistry and Molecular Modeling Principles and Applications*; Springer-Verlag GmbH: New Delhi, India, 2008.

35. Klamt, A. Conductor-like screening model for real solvents: A new approach to the quantitative calculation of solvation phenomena. *J. Phys. Chem. A* **1995**, *99*, 2224–2235.

36. Hukkerikar, A.S.; Sarup, B.; Ten Kate, A.; Abildskov, J.; Sin, G.; Gani, R. Group-contribution$^+$ (GC$^+$) based estimation of properties of pure components: Improved property estimation and uncertainty analysis. *Fluid Phase Equilibr.* **2012**, *321*, 25–43.

37. Papadopoulos, A.I.; Stijepovic, M.; Linke, P.; Seferlis, P.; Voutetakis, S. Multi-level design and selection of optimum working fluids and ORC systems for power and heat cogeneration from low enthalpy renewable sources. *Comput. Aided Chem. Eng.* **2012**, *30*, 66–70.

38. Papadopoulos, A.I.; Linke, P. Multiobjective molecular design for integrated process-solvent systems synthesis. *AIChE J.* **2006**, *52*, 1057–1069.

39. Zyhowski, G.; Brown, A. Low global warming fluids for replacement of HFC-245fa and HFC-134a in ORC applications. In Proceedings of First International Seminar on ORC systems, Delft, The Netherlands, 22–23 September 2011.

40. Schwiegel, M.; Flohr, F.; Meurer, C. Working Fluid for an Organic Rankine Cycle Process, ORC Process and ORC Apparatus. Patent No. US 20110162366, 7 July 2011.

41. Wang, H.; Zhang, S.; Guo, T.; Chen, C. HFO-1234yf-Containing Mixed Working Fluid for Organic Rankine Cycle. Patent No. CN101747867, 23 June 2010.

42. Wang, J.; Zhang, J.; Chen, Z. Molecular entropy, thermal efficiency, and designing of working fluids for organic Rankine cycles. *Int. J. Thermophys.* **2012**, *33*, 970–985.

43. Palma-Flores, O.; Flores-Tlacuahuac, A.; Canseco-Melchor, G. Optimal molecular design of working fluids for sustainable low-temperature energy recovery. *Comput. Chem. Eng.* **2014**, *72*, 334–339.

44. Lampe, M.; Groß, J.; Bardow, A. Simultaneous process and working fluid optimization for organic Rankine cycles (ORC) using PC-SAFT. *Comput. Aided Chem. Eng.* **2012**, *30*, 572–576.

45. Lampe, M.; Stavrou, M.; Bücker, M.; Gross, J.; Bardow, A. Simultaneous optimization of working fluid and process for organic Rankine cycles (ORCs) using PC-SAFT. *Ind. Eng. Chem. Res.* **2014**, *53*, 8821–8830.

46. Roskosch, D.; Atakan, B. Reverse engineering of fluid selection for thermodynamic cycles with cubic equations of state, using a compression heat pump as example. *Energy* **2015**, *81*, 202–212.

47. Samudra, A.P.; Sahinidis, N.V. Optimization-based framework for computer-aided molecular design. *AIChE J.* **2013**, *59*, 3686–3701.

48. Sahinidis, N.V.; Tawarmalani, M.; Yu, M. Design of alternative refrigerants via global optimization. *AIChE J.* **2003**, *49*, 1761–1775.

49. Duvedi, A.P.; Achenie, L.E.K. Designing environmentally safe refrigerants using mathematical programming. *Chem. Eng. Sci.* **1996**, *51*, 3727–3739.

50. Marcoulaki, E.C.; Kokossis, A.C. On the development of novel chemicals using a systematic synthesis approach. Part I. Optimisation framework. *Chem. Eng. Sci.* **2000**, *55*, 2529–2546.

51. Molina-Thierry, D.P.; Flores-Tlacuahuac, A. Simultaneous optimal design of organic mixtures and Rankine cycles for low-temperature energy recovery. *Ind. Eng. Chem. Res.* **2015**, *54*, 3367–3383.

52. Papadopoulos, A.I.; Linke, P. On the synthesis and optimization of liquid-liquid extraction processes using stochastic search methods. *Comput. Chem. Eng.* **2004**, *28*, 2391–2406

53. Cavazzuti, M. *Optimization Methods: From Theory to Design*, Springer-Verlag: Berlin, Germany, 2013.

54. Kasaš, M.; Kravanja, Z.; Novak Pintarič, Z. Suitable modeling for process flow sheet optimization using the correct economic criterion. *Ind. Eng. Chem. Res.* **2011**, *50*, 3356–3370.

55. Harinck, J.; Pasquale, D.; Pecnik, R.; van Buijtenen, J.; Colonna, P. Performance improvement of a radial organic Rankine cycle turbine by means of automated computational fluid dynamic design. *J. Power Energy* **2013**, *227*, 637–645.

56. Weith, T.; Heberle, F.; Preißinger, M.; Brüggemann, D. Performance of siloxane mixtures in a high-temperature organic Rankine cycle considering the heat transfer characteristics during evaporation. *Energies* **2014**, *7*, 5548–5565.

57. Franco, A; Villani, M. Optimal design of binary cycle power plants for water-dominated, medium-temperature geothermal fields. *Geothermics* **2009**, *38*, 379–391.

58. Salcedo, R.; Antipova, E.; Boer, D.; Jimenez, L.; Guillen-Gosalbez, G. Multi-objective optimization of solar Rankine cycles coupled with reverse osmosis desalination considering economic and life cycle environmental impacts. *Desalination* **2012**, *286*, 358–371.

59. Wang, J.; Yan, Z.; Wang, M.; Ma, S.; Dai, Y. Thermodynamic analysis and optimization of an (organic rankine cycle) ORC using low grade heat source. *Energy* **2013**, *49*, 356–365.

60. Wang, J.; Yan, Z.; Wang, M.; Li, M.; Dai, Y. Multi-objective optimization of an organic Rankine cycle (ORC) for low grade waste heat recovery using evolutionary algorithm. *Energy Convers. Manag.* **2013**, *71*, 146–158.

61. Wang, J.; Wang, M.; Li, M.; Xia, J.; Dai, Y. Multi-objective optimization design of a condenser in an organic Rankine cycle for low grade waste heat recovery using evolutionary algorithm. *Int. Commun. Heat Mass* **2013**, *45*, 47–54.

62. Xi, H.; Li, N.J.; Xu, C.; He, Y.L. Parametric optimization of regenerative organic Rankine cycle (ORC) for low grade waste heat recovery using genetic algorithm. *Energy* **2013**, *58*, 473–482.

63. Walraven, D.; Laenen, B.; D'haeseleer, W. Economic system optimization of air cooled organic Rankine cycles powered by low temperature geothermal heat sources. *Energy* **2015**, *80*, 104–113

64. Walraven, D.; Laenen, B.; D'haeseleer, W. Minimizing the levelized cost of electricity production from low temperature geothermal heat sources with ORCs: Water or air cooled? *Appl. Energy* **2015**, *142*, 144–153.

65. Gerber, L.; Marechal, F. Defining optimal configurations for geothermal systems using process design and process integration techniques. *Appl. Therm. Eng.* **2012**, *43*, 29–41.

66. Halemane, K.P.; Grossmann, I.E. Optimal process design under uncertainty. *AIChE J.* **1983**, *29*, 425–433.

67. Pierobon, L.; Nguyen, T.V.; Larsen, U.; Haglind, F.; Elmegaard, B. Multi-objective optimization of organic Rankine cycles in an offshore platform. *Energy* **2013**, *58*, 538–549.

68. Larsen, U.; Pierobon, L.; Haglind, F.; Gabrieli, C. Design and optimisation of organic Rankine cycles for waste heat recovery on marine applications using principles of natural selection. *Energy* **2014**, *55*, 803–812.

69. Clarke, J.; McLeskey, J.T. Multi-objective particle swarm optimization of binary geothermal power plants. *Appl. Energy* **2015**, *138*, 302–314.

70. Stijepovic, M.Z.; Papadopoulos, A.I.; Linke, P.; Grujic, A.S.; Seferlis, P. An exergy Composite curves approach for the design of optimum multi-pressure organic Rankine cycle processes. *Energy* **2014**, *69*, 285–298.

71. Linnhoff, D.; Dhole, V.R. Shaftwork targets for low-temperature process design. *Chem. Eng. Sci.* **1992**, *47*, 2081–2091.

72. Toffolo, A. A synthesis/design optimization algorithm for Rankine cycle based energy systems. *Energy* **2014**, *66*, 115–127.

73. Lecompte, S.; Huisseune, H.; van den Broek, M.; Vanslambrouck, B.; de Paepe, M. Review of organic Rankine cycle (ORC) architectures for waste heat recovery. *Renew. Sustain. Energy Rev.* **2015**, *47*, 448–461.

74. Pardalos, P.M.; Romeijn, H.E.; Tuy, H. Recent developments and trends in global optimization. *J. Comput. Appl. Math.* **2000**, *124*, 209–228.

75. Floudas, C.A.; Akrotirianakis, I.G.; Caratzoulas, S.; Meyer, C.A.; Kallrath, J. Global optimization in the 21st century: Advances and challenges. *Comput. Chem. Eng.* **2005**, *29*, 1185–1202.

76. Floudas, C.A.; Gounaris, C.E. A review of recent advances in global optimization. *J. Glob. Optim.* **2009**, *45*, 3–38.

77. Novak Pintarič, Z.; Kravanja, Z. Selection of the economic objective function for the optimization of process flow sheets. *Ind. Eng. Chem. Res.* **2006**, *45*, 4222–4232.

78. Novak Pintarič, Z.; Kravanja, Z. The importance of proper economic criteria and process modeling for single- and multi-objective optimizations. *Comput. Chem. Eng.* **2015**, doi:10.1016/j.compchemeng.2015.02.008.

79. Marlin, T.E. *Process Control: Designing Processes and Control Systems for Dynamic Performance*; McGraw-Hill: Singapore, 1995.

80. Goodwin, G.C.; Graebe, S.E.; Salgado, M.E. *Control System Design*; Prentice Hall: Upper Saddle River, NJ, USA, 2001.

81. Rossiter, J.A. *Model Predictive Control: A Practical Approach*; CRC Press: Boca Raton, FL, USA, 2003.

82. Bemporad, A.; Morari, M.; Dua, V.; Pistikopoulos, E.N. The explicit linear quadratic regulator for constrained systems. *Automatica* **2002**, *38*, 3–20.

83. Kouvaritakis, B.; Cannon, M. *Nonlinear Predictive Control: Theory and Practice*; The Institution of Engineering and Technology: London, UK, 2001.

84. Zavala, V.M.; Biegler, L.T. The advanced-step NMPC controller: Optimality, stability and robustness. *Automatica* **2009**, *45*, 86–93.

85. Quoilin, S.; Lemort, V.; Lebrun, J. Experimental study and modeling of an organic Rankine cycle using scroll expander. *Appl. Energy* **2010**, *87*, 1260–1268.

86. Smith, C.A.; Corripio, A. *Principles and Practice of Automatic Process Control*; John Wiley & Sons Inc.: Hoboken, NJ, USA, 2006.

87. Box, G.E.P.; Jenkins, G.M.; Reinsel, G.C. *Time Series Analysis: Forecasting and Control*; John Wiley & Sons Inc.: Hoboken, NJ, USA, 2008.

88. Zhang, J.; Zhang, W.; Hou, G.; Fang, F. Dynamic modeling and multivariable control of organic Rankine cycles in waste heat utilizing processes. *Comput. Math. Appl.* **2012**, *64*, 908–921.

89. Zhang, J.; Zhou, Y.; Li, Y.; Hou, G.; Fang, F. Generalized predictive control applied in waste heat recovery power plants. *Appl. Energy* **2013**, *102*, 320–326.

90. Zhang, J.; Zhou, Y.; Wang, R.; Xu, J.; Fang, F. Modeling and constrained multivariable predictive control for ORC (Organic Rankine Cycle) based waste heat energy conversion systems. *Energy* **2014**, *66*, 128–138.

91. Twomey, B.; Jacobs, P.A.; Gurgenci, H. Dynamic performance estimation of small-scale solar cogeneration with an organic Rankine cycle using a scroll expander. *Appl. Therm. Eng.* **2013**, *51*, 1307–1316.

92. Wei, D.; Lu, X.; Lu, Z.; Gu, J. Dynamic modeling and simulation of an organic Rankine cycle (ORC) system for waste heat recovery. *Appl. Therm. Eng.* **2008**, *28*, 1216–1224.

93. Bamgbopa, M.O.; Uzgoren, E. Quasi-dynamic model for an organic Rankine cycle. *Energy Convers. Manag.* **2013**, *72*, 117–124.

94. Bamgbopa, M.O.; Uzgoren, E. Numerical analysis of an organic Rankine cycle under steady and variable heat input. *Appl. Energy* **2013**, *107*, 219–228.

95. Quoilin, S.; Aumann, R.; Grill, A.; Schuster, A.; Lemort, V.; Spliethoff, H. Dynamic modeling and optimal control strategy of waste heat recovery organic Rankine cycles. *Appl. Energy* **2013**, *88*, 2183–2190.

96. Peralez, J.; Tona, P.; Lepreux, O.; Sciarretta, A.; Voise, L.; Dufour, P.; Nadri, M. Improving the control performance of an organic Rankine cycle system for waste heat recovery from a heavy-duty diesel engine using a model-based approach. In Proceedings of the 2013 IEEE Conference on Decision and Control (CDC), Florence, Italy, 10–13 December 2013.

97. Kosmadakis, G.; Manolakos, D.; Papadakis, G. An investigation of design concepts and control strategies of a double-stage expansion solar organic Rankine cycle. *Int. J. Sustain. Energy* **2015**, *34*, 446–467.

98. Luong, D.; Tsao, T.-C. Linear quadratic integral control of an organic Rankine cycle for waste heat recovery in heavy-duty diesel powertrain. In Proceedings of the American Control Conference (ACC), Portland, OR, USA, 4–6 June 2014; pp. 3147–3152.

99. Zhang, J.; Feng, J.; Zhou, Y.; Fang, F.; Yue, H. Linear active disturbance rejection control of waste heat recovery systems with organic Rankine cycles. *Energies* **2012**, *5*, 5111–5125.

100. Camacho, E.F.; Bordons, C. *Model Predictive Control*; Springer-Verlag: London, UK, 2004.

101. Hou, G.; Bi, S.; Lin, M.; Zhang, J.; Xu, J. Minimum variance control of organic Rankine cycle based waste heat recovery. *Energy Convers. Manag.* **2014**, *86*, 576–586.

102. Ibarra, M.; Rovira, A.; Alarcón-Padilla, D.C.; Blanco, J. Performance of a 5kWe organic Rankine cycle at part-load operation. *Appl. Energy* **2014**, *120*, 147–158.

103. Manente, G.; Toffolo, A.; Lazzaretto, A.; Paci, M. An organic Rankine cycle off-design model for the search of the optimal control strategy. *Energy* **2013**, *58*, 97–106.

104. De Escalona, J.M.; Sánchez, D.; Chacartegui, R.; Sánchez, T. Part-load analysis of gas turbine & ORC combined cycles. *Appl. Therm. Eng.* **2012**, *36*, 63–72.

105. Pierobon, L.; Nguyen, T.V.; Mazzucco, A.; Larsen, U.; Haglind, F. Part load performance of a wet indirectly fired gas turbine integrated with an ORC turbogenerator. *Energies* **2014**, *7*, 8294–8316.

106. Stijepovic, M.Z.; Linke, P. Optimal waste heat recovery and reuse in industrial zones. *Energy* **2011**, *36*, 4019–4031

107. Romeo, L.M.; Lara, Y.; Gonzalez, A. Reducing energy penalties in carbon capture with organic Rankine cycles. *Appl. Therm. Eng.* **2011**, *31*, 2928–2935.

108. Soffiato, M.; Frangopoulos, C.A.; Manente, G.; Rech, S.; Lazzaretto, A. Design optimization of ORC systems for waste heat recovery on board a LNG carrier. *Energy Convers. Manag.* **2015**, *92*, 523–534.

109. Linhoff, B.; Flower, J.R. Synthesis of heat exchanger networks: I. Systematic generation of energy optimal networks. *AIChE J.* **1978**, *24*, 633–642.

110. Smith, R. *Chemical Process: Design and Integration*; John Wiley and Sons: New York, NY, USA, 2005.

111. Varbanov, P.S.; Fodor, Z.; Klemeš, J.J. Total Site targeting with process specific minimum temperature difference (ΔTmin). *Energy* **2012**, *44*, 20–28.

112. Hackl, R.; Harvey, S. Applying process integration methods to target for electricity production from industrial waste heat using Organic Rankine Cycle (ORC) technology. In Proceedings of the World Renewable Energy Congress, Linköping, Sweden, 8–11 May 2011.

113. Desai, B.D.; Bandyopadhyay, S. Process integration of organic Rankine cycle. *Energy* **2009**, *34*, 1674–1686.

114. Kapil, A.; Bulatov, I.; Smith, R.; Kim, J.K. Site-wide low-grade heat recovery with a new cogeneration targeting method. *Chem. Eng. Res. Des.* **2012**, *90*, 677–689.

115. Hipólito-Valencia, B.J.; Rubio-Castro, E.; Ponce-Ortega, J.M.; Serna-González, M.; Nápoles-Rivera, F.; El-Halwagi, M.M. Optimal integration of organic Rankine cycles with industrial processes. *Energy Convers. Manag.* **2013**, *73*, 285–302.

116. Hipólito-Valencia, B.J.; Rubio-Castro, E.; Ponce-Ortega, J.M.; Serna-González, M.; Nápoles-Rivera, F.; El-Halwagi, M.M. Optimal design of inter-plant waste energy integration. *Appl. Therm. Eng.* **2014**, *62*, 633–652.

117. Lira-Barragán, L.F.; Ponce-Ortega, J.M.; Serna-González, M.; El-Halwagi, M.M. Sustainable integration of trigeneration systems with heat exchanger networks. *Ind. Eng. Chem. Res.* **2014**, *53*, 2732–2750.

118. Gutiérrez-Arriaga, C.G.; Abdelhady, F.; Bamufleh, H.S.; Serna-González, M.; El-Halwagi, M.M.; Ponce-Ortega, J.M. Industrial waste heat recovery and cogeneration involving organic Rankine cycles. *Clean Technol. Environ. Policy* **2015**, *17*, 767–779.

119. Kwak, D.-H.; Binns, M.; Kim, J.-K. Integrated design and optimization of technologies for utilizing low grade heat in process industries. *Appl. Energy* **2014**, *131*, 307–322.

120. Chen, C.-L.; Chang, F.-Y.; Chao, T.-H.; Chen, H.-C.; Lee, J.-Y. Heat-Exchanger network synthesis involving organic Rankine cycle for waste heat recovery. *Ind. Eng. Chem. Res.* **2014**, *53*, 16924–16936.

121. Marechal, F.; Kalitventzeff, B. A methodology for the optimal insertion of organic Rankine cycles in industrial processes. In *2nd International Symposium of Process Integration*; Dalhousie University, Halifax, Canada, 2004.

122. Song, J.; Li, Y.; Gu, C.W.; Zhang, L. Thermodynamic analysis and performance optimization of an ORC (Organic Rankine Cycle) system for multi-strand waste heat sources in petroleum refining industry. *Energy* **2014**, *71*, 673–680.

123. Tchanche, B.F.; Lambrinos, G.; Frangoudakis, A.; Papadakis, G. Exergy analysis of micro-organic Rankine power cycles for a small scale solar driven reverse osmosis desalination system. *Appl. Energy* **2010**, *87*, 1295–1306.

124. Yu, H.; Feng, X.; Wang, Y. A new pinch based method for simultaneous selection of working fluid and operating conditions in an ORC (Organic Rankine Cycle) recovering waste heat. *Energy* **2015**, doi:10.1016/j.energy.2015.02.059.

125. Safarian, S.; Aramoun, F. Energy and exergy assessments of modified Organic Rankine Cycles (ORCs). *Energy Rep.* **2015**, *1*, 1–7.

126. Luo, X.; Zhang, B.; Chen, Y.; Mo, S. Heat integration of regenerative Rankine cycle and process surplus heat through graphical targeting and mathematical modeling technique. *Energy* **2012**, *45*, 556–569.

6

Molecular and Isotopic Composition of Volatiles in Gas Hydrates and in Sediment from the Joetsu Basin, Eastern Margin of the Japan Sea

Akihiro Hachikubo [1,*], Katsunori Yanagawa [2], Hitoshi Tomaru [3], Hailong Lu [4] and Ryo Matsumoto [5]

[1] Environmental and Energy Resources Research Center, Kitami Institute of Technology, 165 Koen-cho, Kitami 090-8507, Japan

[2] Faculty of Social and Cultural Studies, Kyushu University, 744 Motooka, Nishi-ku, Fukuoka 819-0395, Japan; E-Mail: kyanagawa@scs.kyushu-u.ac.jp

[3] Department of Earth Sciences, Graduate School of Science, Chiba University, 1-33 Yayoi-cho, Inage-ku, Chiba 263-8522, Japan; E-Mail: tomaru@chiba-u.jp

[4] Department of Energy and Resource Engineering, College of Engineering, Peking University, Beijing 100871, China; E-Mail: hlu@pku.edu.cn

[5] Gas Hydrate Laboratory, Organization for the Strategic Coordination of Research and Intellectual Properties, Meiji University, 1-1 Kanda-Surugadai, Chiyoda-ku, Tokyo 101-8301, Japan; E-Mail: ryo_mat@meiji.ac.jp

* Author to whom correspondence should be addressed; E-Mail: hachi@mail.kitami-it.ac.jp;

Academic Editor: Richard B. Coffin

Abstract: Hydrate-bearing sediment cores were retrieved from the Joetsu Basin (off Joetsu city, Niigata Prefecture) at the eastern margin of the Japan Sea during the MD179 gas hydrates cruise onboard R/V *Marion Dufresne* in June 2010. We measured molecular and stable isotope compositions of volatiles bound in the gas hydrates and headspace gases obtained from sediments to clarify how the minor components of hydrocarbons affects to gas hydrate crystals. The hydrate-bound hydrocarbons at Umitaka Spur (southwestern Joetsu Basin) primarily consisted of thermogenic methane, whereas those at Joetsu Knoll (northwestern Joetsu Basin, about 15 km from Umitaka Spur) contained both thermogenic methane and a mixture of thermogenic and microbial methane. The depth concentration profiles of methane, ethane, propane, CO_2, and H_2S in the sediments from the Joetsu Basin area showed shallow sulfate–methane interface (SMI) and high microbial methane production beneath the SMI depth. Relatively high concentrations of propane and neopentane (2,2-dimethylpropane) were detected in the headspace gases of the hydrate-bearing sediment cores obtained at Umitaka Spur and Joetsu Knoll. Propane and neopentane cannot be encaged in the structure I hydrate; therefore, they were probably excluded from the hydrate crystals during the structure I formation process and thus remained in the sediment and/or released from the small amounts of structure II hydrate that can host such large gas molecules. The lower concentrations of ethane and propane in the sediment, high $\delta^{13}C$ of propane and isobutane, and below-detection normal butane and normal pentane at Umitaka Spur and Joetsu Knoll suggest biodegradation in the sediment layers.

Keywords: gas hydrate; Japan Sea; stable isotope; neopentane; biodegradation

1. Introduction

Gas hydrates are crystalline clathrate compounds consisting of water and gas molecules that form at low temperatures and high pressures [1]. Natural gas hydrates are found worldwide in continental margin sediments [2–4] and in near-surface sediments associated with active gas plume that vent from the seafloor [5–7]. Natural gas hydrates are considered potential energy resources and are large reservoirs of methane (C_1), and their dissociation may cause submarine geohazards and contribute to global warming [8–10].

Crystallographic structures of natural gas hydrate are usually either structure I (sI), which is composed of two 12-hedra and six 14-hedra with space group Pm3n, or structure II (sII), which is composed of sixteen 12-hedra and eight 16-hedra with space group Fd3m [1]. C_1 and ethane (C_2) are both known to form sI hydrates; however, certain compositions of C_1 and C_2 will form sII hydrates [11,12], whereas propane (C_3) and isobutane (i-C_4) is incorporated only in sII. The concentrations of C_2 and C_3 in hydrate-bound gas from the Gulf of Mexico represent 3%–5% and >15% of the total gas, respectively [13]. Gas hydrate from the Sea of Marmara contained high concentrations of C_3 (18.8%) and i-C_4 (9.5%) [14]. Mixed-gas (C_1 and C_2) hydrates in Lake Baikal belonged to the sII hydrate and contained 0.026%–0.064% of neopentane (neo-C_5, 2,2-dimethylpropane) [15,16], which can be encaged in the large cages of sII [17].

Shallow gas hydrates were recently found in the Joetsu Basin at the eastern margin of the Japan Sea, where gas venting was observed on echo-sounder images [18,19]. Gas hydrate was recovered from the sea floor [20], and $\delta^{13}C$ of hydrate-bound C_1 was from −37.3‰ to −37.1‰ [20], suggesting thermogenic origin in the criteria of the Bernard plot [21]. However, C_2 and C_3 concentrations of hydrate-bound gas at Umitaka Spur and Joetsu Knoll (Figure 1) were very low; the molar ratio $C_1/(C_2 + C_3)$ ranged from 1000 to 10,000 [22]. The Sado Nansei Oki drillings at Umitaka Spur conducted by the Ministry of Economy, Trade and Industry, Japan revealed $C_1/(C_2 + C_3)$ values less than 100 at depths 1143−2016 m below the sea floor [20], indicating either that C_2 and C_3 concentrations are reduced in the ascending fluid during migration or that the addition of microbial C_1 is significant in higher strata. The depth profiles of sediment gases at Umitaka Spur and Joetsu Knoll were reported in our previous work [23]; however, molecular and isotopic compositions of hydrate-bound gas were not reported yet and the effect of low concentration of higher hydrocarbons on gas hydrates has not been discussed. In this study, we investigate the molecular and stable isotope compositions of hydrate-bound gas and gas in sediment cores (headspace gas) retrieved from Umitaka Spur and Joetsu Knoll and focus on C_2, C_3, and higher hydrocarbons, those may change crystallographic structure of the shallow gas hydrates in these areas.

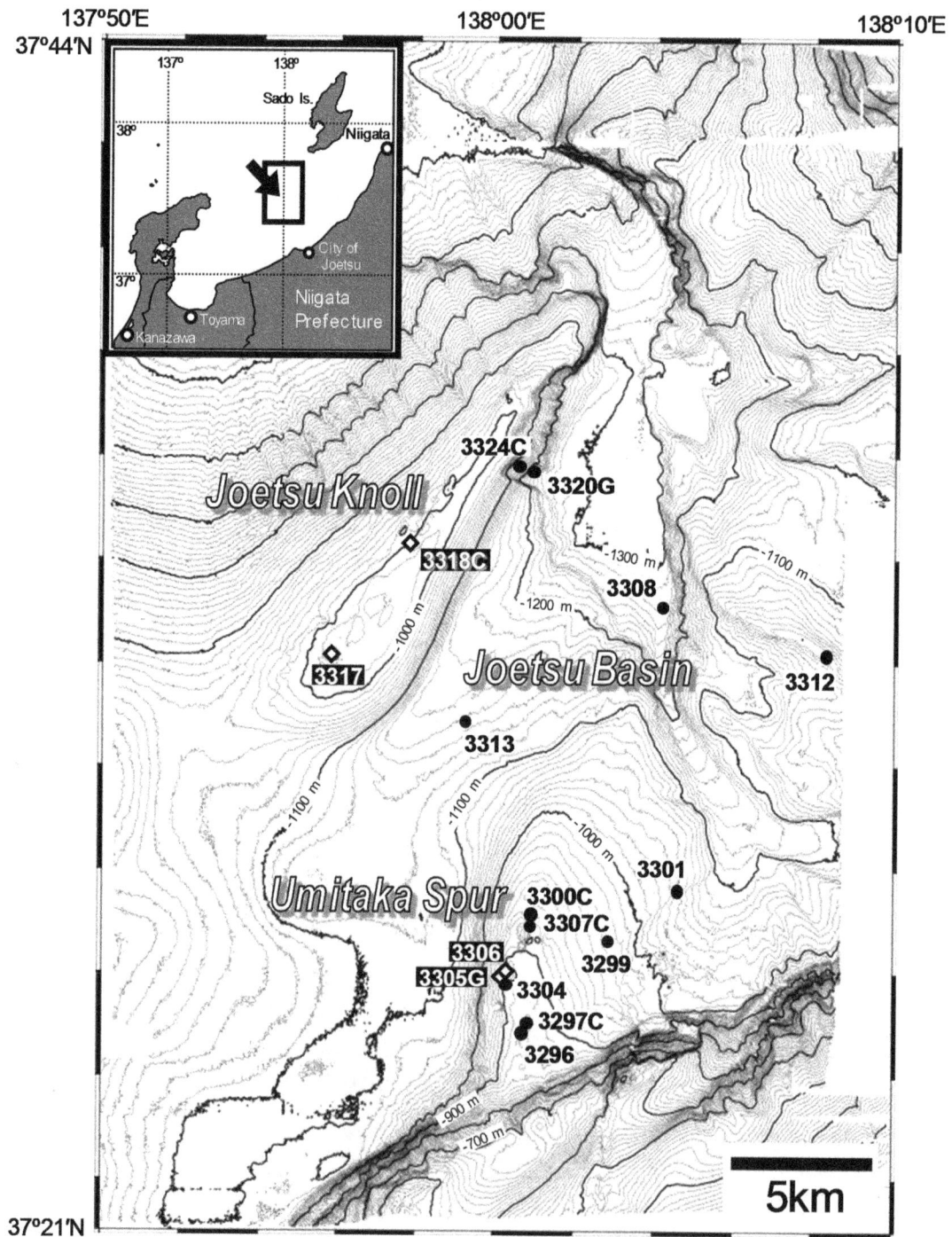

Figure 1. Locations of the coring sites in the Joetsu Basin area, Japan Sea. The sites of hydrate-bearing cores are highlighted.

2. Study Sites and Sediment Cores

Umitaka Spur and Joetsu Knoll are located in the Joetsu Basin off the city of Joetsu (Niigata Prefecture, Figure 1). High P-wave velocities at the mounds and pockmarks on the Umitaka Spur seafloor were reported, suggesting the existence of gas hydrate [24]. Echo sounding of the sea bottom at Umitaka Spur and Joetsu Knoll showed large-scale gas venting from the sea floor [18]. Depth profiles of sulfate in the pore water of the sediment cores indicated shallow (<5 m) sulfate–methane interface (SMI; where sulfate and methane are both consumed to depletion at the base of the sulfate reduction zone) in the Joetsu Basin [25], indicating high flux of hydrocarbons and other organic matter. High heat flow was observed in a restricted area on the sea floor around the gas seep sites at Umitaka Spur (maximum value of 1590 mW m^{-2}) and Joetsu Knoll (maximum value of 519 mW m^{-2}) compared to the average value obtained on a normal muddy seafloor in these areas (98 ± 13 mW m^{-2}), indicating local fluid migration from the deep subsurface layer [26].

Previous study considered origin of hydrate-bound C_1 at Umitaka Spur and Joetsu Knoll as thermogenic one (C_1 $\delta^{13}C$ ranged from −39.0 to −34.9‰ V-PDB) and mixed gas of microbial and thermogenic ones (C_1 $\delta^{13}C$ ranged from −57.9 to −51.8‰ V-PDB), respectively [22]. Because the content of total organic carbon (TOC) ranged from 0.58% to 1.55% in the shallow sediment layer (0–9 m depth, [27]) and from 0.90% to 2.98% in the deep layer (1360–2088 m depth, mixed Type II/III kerogen, [20]), the potential of C_1 generation via microbial and thermogenic processes is high.

Sediment cores were retrieved from Umitaka Spur, Joetsu Knoll, Joetsu Basin and its peripheral area (Figure 1) during the MD179 cruise onboard R/V *Marion Dufresne* in June 2010. A Calypso giant piston corer enabled us to obtain sediment cores up to 40 m in length. A Calypso square box corer (CASQ, 25 cm × 25 cm × 12 m) was also used. Out of the 20 sediment cores obtained, four contained hydrates: MD179-3305G and MD179-3306 (located 300 m apart) from Umitaka Spur, and MD179-3317 and MD179-3318C (located 5 km apart) from Joetsu Knoll. Hydrates were virtually absent in the other sediment cores. MD179-3305G was retrieved by a gravity corer, and a few samples of the gas hydrates were taken for gas analysis. All the hydrate-bearing cores except MD179-3318C lost several meters of the top sediment layers because the lead weight section was disconnected from the core barrel due to a technical problem; the gas hydrates dissociated in the corer and the core top was displaced during the onboard recovery process. The top depths of the gas hydrates were estimated to be 4.5 mbsf (MD179-3306), 31.5 mbsf (MD179-3317), and 1.5 mbsf (MD179-3318C). The depth of gas hydrate in MD179-3305G was unknown.

3. Sampling Methods and Analysis

3.1. Onboard Gas Sampling

The gas sampling procedures for the hydrate-bound and headspace gases were the same as those used in our previous studies [28,29]. The hydrate-bound gases were collected using a 50-mL plastic syringe and stored in 5-mL vials sealed with butyl septum stoppers. Several samples were taken from each hydrate-bearing sediment core. First, we placed small pieces of hydrate sample into a 50-mL plastic syringe, pushed the cylinder to reduce the dead volume, and connected the syringe to a 5-mL vial with a needle. Another needle was connected to the vial to flush the air inside. Each vial was thus filled with hydrate-bound gas without sediment particles or water. Consequently, this method contributed to prevention of microbial activity during storage. The gas dissociated from hydrate flushed air in the system, and the concentration of air in the vial was less than 2%.

Headspace gas method was employed to know the depth profiles of each gas component in the sediment cores. 10 mL of sediment, 9.5 mL of saturated aqueous solution of NaCl, and 0.5 mL of preservative (10 wt % aqueous solution of benzalkonium chloride, [30]) were introduced into 25-mL vials to create a 5-mL headspace. The headspace was flushed with helium, the carrier gas used in the gas chromatography, to reduce air contamination. Although the flushing process may reduce gas concentrations from their native values, differences in concentration and stable isotopes ($\delta^{13}C$ and δD) of C_1 between flushed and non-flushed samples were undetectable [31]. Hence, we can discuss the trend of their depth profiles using headspace gas data. The vials were shaken well and stored upside down until they were analyzed.

In some of the gas-rich cores (MD179-3296, 3299, 3301, 3304, 3308, 3313, and 3317, see Figure 1), voids formed in the plastic core liner during their retrieval from the core barrel. The gas in the voids was collected by modifying the method of [32]. The surface temperature of the plastic core liner was measured with an infrared thermal imaging camera to locate the voids. Small holes several meters apart were then drilled into the liner. Gas samples collected through these holes with a 50-mL plastic syringe were stored in evacuated 5-mL glass vials. Because the concentration of hydrocarbons in the void gas was considerably higher than that obtained by the headspace gas method, the void-gas method is suitable for stable carbon and hydrogen isotope analysis on non-C_1 hydrocarbons, which are typically present in low concentration. Isotopic difference between void and headspace gases was within 2‰ for C_1 $\delta^{13}C$ and 3‰ for C_1 δD [23]. In this study, we obtained the stable carbon isotope signature of C_2 by the void-gas method.

3.2. Analytical Methods

The molecular composition of the gas samples was determined using a gas chromatograph (GC-14B, Shimadzu, Kyoto, Japan) equipped with a packed column (Shimadzu Sunpak-S; 2 m length, 3 mm ID), a thermal conductivity detector (TCD) for detecting CO_2, H_2S, and high concentrations (>0.1 mol% of the sample gas) of C_1, and a flame ionization detector (FID) for detecting low concentrations (<0.1 mol% of the sample gas) of hydrocarbons (C_1–C_5). The TCD and FID were connected in series. The detection limit was 0.00005 mol% (C_1–C_3) and 0.0005 mol% (C_4–C_5). The analytical error estimated by multiple injections of standard gases was less than 1.2% for each gas component.

Stable carbon and hydrogen isotopic ratios of the hydrocarbons and CO_2 were measured using a continuous-flow isotope-ratio mass spectrometer (CF-IRMS, DELTAplus XP, Thermo Fisher Scientific, Waltham, MA, USA) coupled with a gas chromatograph (Trace GC, Thermo Fisher Scientific) via a combustion/pyrolysis reactor (GC-C/TC III, Thermo Fisher Scientific). The gas chromatograph was equipped with a Carboxen-1006 PLOT capillary column (30 m length, 0.32 mm ID, 15 μm film thickness, Sigma-Aldrich, St. Louis, MO, USA). In the case of samples with low C_1 concentration, a Sigma-Aldrich Carboxen-1010 PLOT capillary column (30 m length, 0.32 mm ID, 15 μm film thickness) was also used to separate air components from hydrocarbons. A PoraPLOT Q capillary column

(27.5 m length, 0.32 mm ID, 10 μm film thickness, Agilent Technologies, Santa Clara, CA, USA) was used for higher hydrocarbons (C_4–C_8). Stable isotope compositions are reported as δ values (‰):

$$\delta_{sample} = \left(\frac{R_{sample} - R_{standard}}{R_{standard}} \right) \times 1000 \ (‰) \tag{1}$$

where R denotes the $^{13}C/^{12}C$ or D/H ratio. $\delta^{13}C$ and δD are given with reference to the V-PDB and V-SMOW standards, respectively, determined by using NIST RM8544 (NBS19) for $\delta^{13}C$ and NIST RM8561 (NGS3) for δD. The analytical precision was 0.3‰ for $\delta^{13}C$ of C_1–C_3, 1‰ for $\delta^{13}C$ of C_4–C_8, and 2‰ for δD.

4. Results

Figure 2a–c show the depth profiles of the concentration (C_1, C_2, C_3, CO_2, and H_2S) of headspace gases, $C_1/(C_2 + C_3)$ values calculated from the concentration of C_1, C_2, and C_3, and stable carbon and hydrogen isotopes of C_1, C_2, and CO_2. The concentration of C_1 drastically increased with depth toward the SMI, with the peak concentration appearing at around 5–8 mbsf and then decreasing slightly. The C_1 concentration profiles indicate that the SMI is around 0.5–5 mbsf in the Joetsu Basin area including Umitaka Spur and Joetsu Knoll. The trend of C_2 and C_3 concentration profiles was similar to that of C_1; however, their concentrations below the SMI differed between the sediment cores. $C_1/(C_2 + C_3)$ values showed a peak near 5–8 m, which agrees with the peak of the C_1 concentrations. $C_1/(C_2 + C_3)$ values below the SMI were less than 10,000 in the areas of Umitaka Spur (Figure 2a) and Joetsu Knoll (Figure 2b) except the cores MD179-3320G and 3324C, and more than 10,000 in the other cores of Joetsu Basin (Figure 2c). The CO_2 concentrations of the all cores simply increased with depth, and the H_2S concentrations showed peaks at their SMI depths.

The trend of C_1 $\delta^{13}C$ profiles were similar to those of CO_2 $\delta^{13}C$, increasing with depth below the SMI. Negative C_1 $\delta^{13}C$ and CO_2 $\delta^{13}C$ peaks were observed around the SMI, which agreed with the high H_2S concentrations. High C_1 $\delta^{13}C$ (more than −45‰ V-PDB) was observed in the hydrate-bearing cores MD179-3306 and 3318C. C_1 $\delta^{13}C$ at the depth of 10 mbsf ranged from −66.7‰ to −57.8‰ at Umitaka Spur except the core MD-179-3306, from −74.5‰ to −58.3‰ at Joetsu Knoll, and from −83.5‰ to −77.4‰ at the other sites in Joetsu Basin. On the other hand, C_1 δD simply decreased with depth except the core MD179-3306. C_1 δD for MD179-3306 was distinctly large and ranged from −165.9‰ to −157.8‰, while those for other cores ranged from −207.4‰ to −175.8‰. C_1 $\delta^{13}C$, C_1 δD, and C_2 $\delta^{13}C$ of hydrate-bound and headspace gases were almost the same with each other. On the contrary, distinct differences were found in CO_2 $\delta^{13}C$; those of hydrate-bound gas was 10‰ lower (MD179-3306), 8‰ lower (MD179-3317), and 27‰ higher (MD179-3318C) than those of headspace gas around the same depth.

The concentration of C_2 in the hydrate-bearing cores (MD179-3306, 3317, and 3318C) below the SMI was nearly one order of magnitude higher than that in the nonhydrate-bearing cores. MD179-3297C, 3300C, 3304, and 3307C were also rich in C_2. The concentrations of C_3 in the three hydrate-bearing cores (MD179-3306, 3317, and 3318C) and in MD179-3304 were higher than in the nonhydrate-bearing cores. The C_2-rich cores (MD179-3304 and 3317) showed ^{13}C enrichment in C_2, although C_3 $\delta^{13}C$ is not available. C_2 $\delta^{13}C$ of nonhydrate-bearing cores in Joetsu Basin was low, ranging from −62.8‰ to −52.6‰.

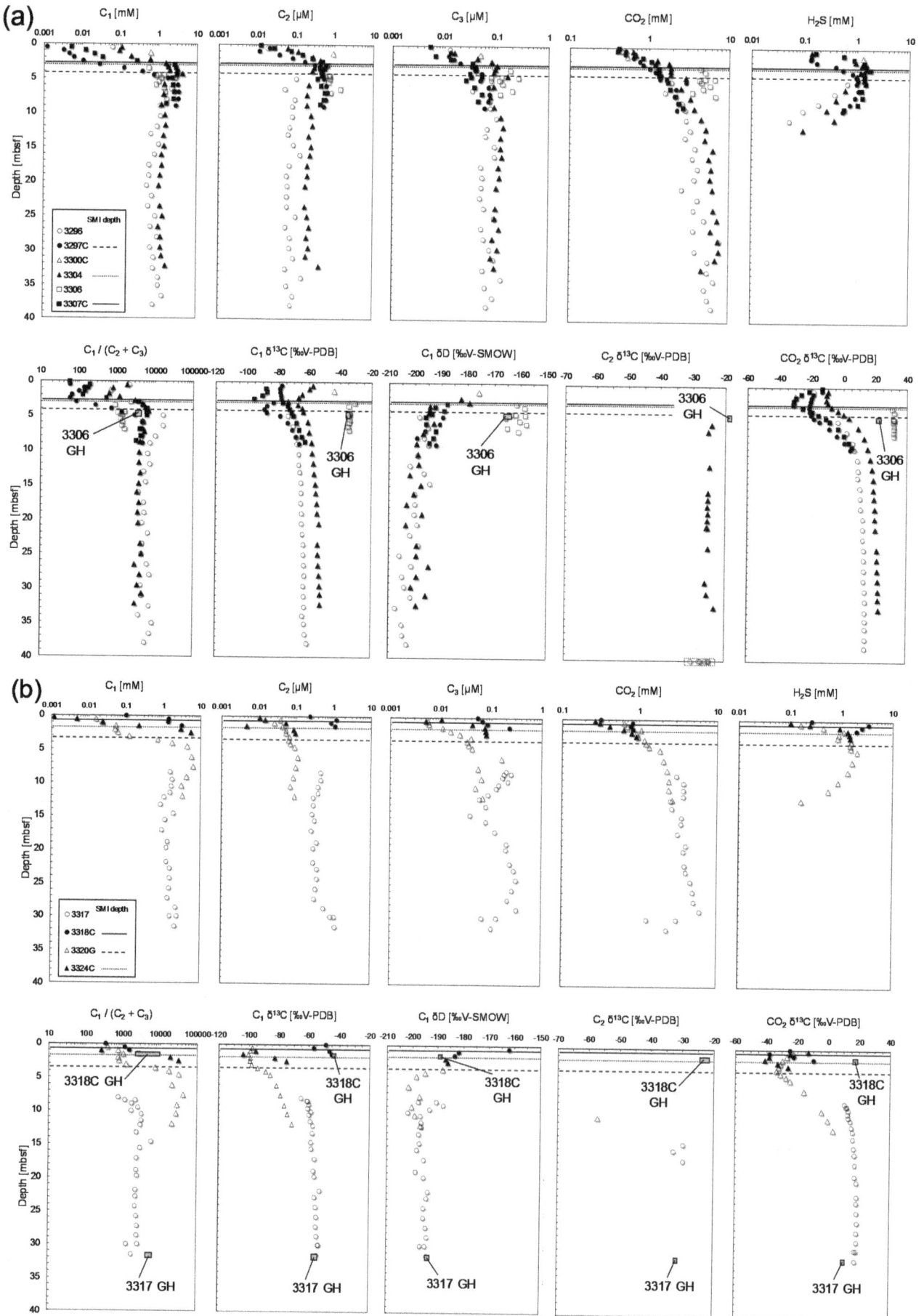

Figure 2. *Cont.*

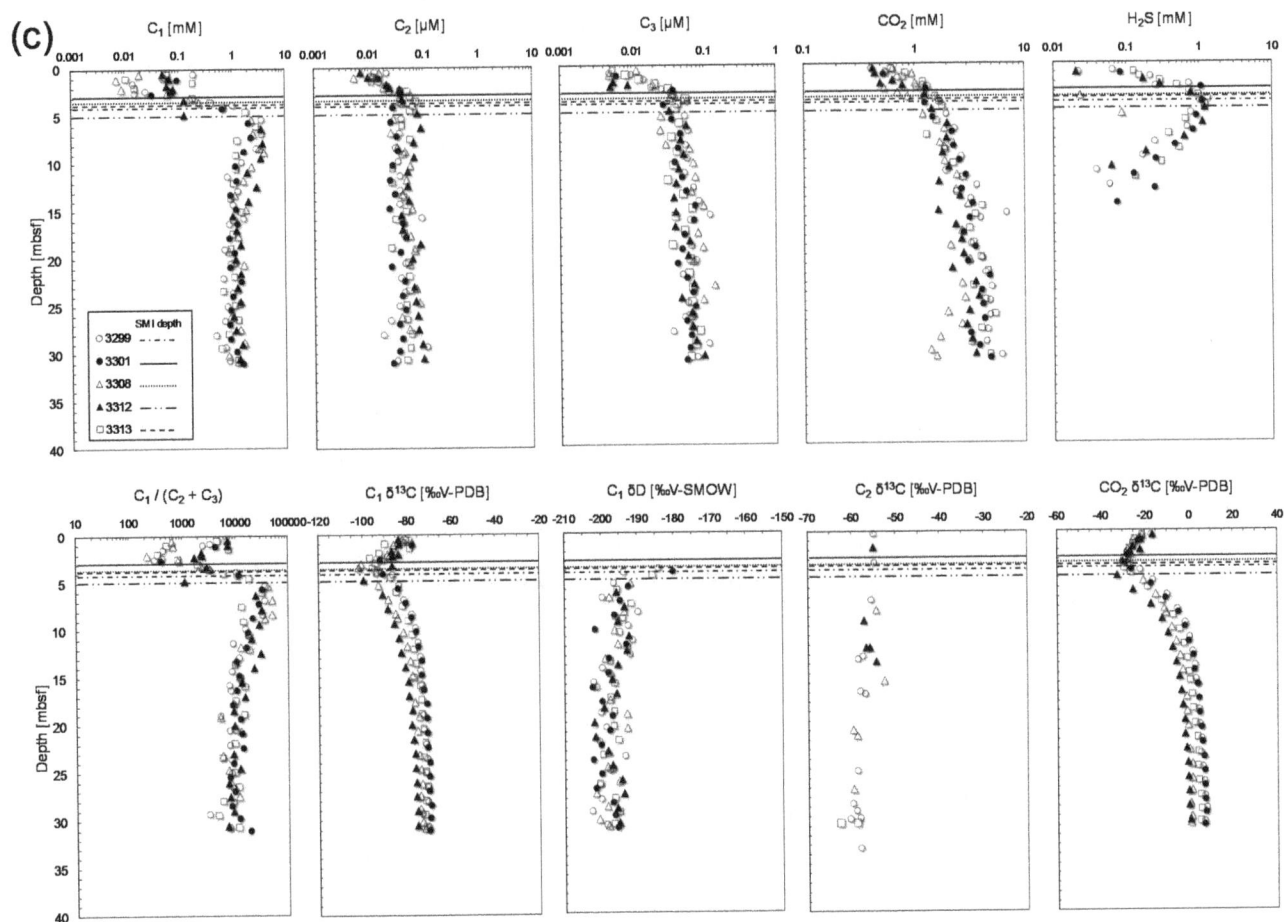

Figure 2. Depth profiles of C_1, C_2, C_3, CO_2, and H_2S concentrations, $C_1/(C_2 + C_3)$ values, C_1 $\delta^{13}C$, C_1 δD, C_2 $\delta^{13}C$, and CO_2 $\delta^{13}C$ in the headspace gas. (**a**) Umitaka Spur area; (**b**) Joetsu Knoll area; (**c**) Joetsu Basin and its peripheral area. These depth profiles of the headspace gases were already reported [23]. The $C_1/(C_2 + C_3)$ values and stable carbon (C_1, C_2, and CO_2) and hydrogen (C_1) isotopic ratios in the hydrate-bound gas of the cores MD179-3306, 3317 and 3318C are plotted. The data of C_2 $\delta^{13}C$ was obtained by the void-gas method and other data by the headspace gas method. The depth data of C_2 $\delta^{13}C$ profile for MD179-3296 is unknown. The headspace gas data of MD179-3305G is missing due to sediment loss.

Molecular compositions and stable isotope signatures of hydrate-bound gases at Umitaka Spur and Joetsu Knoll are summarized in Table 1. In the entire study sites C_1 was the main component of the hydrate-bound gas, comprising more than 98 mol% of the total volume; the concentrations of C_2, C_3, CO_2, and H_2S were 0.0148–0.0456 mol% ($n = 17$), 0.0001–0.0025 mol% ($n = 17$), 0.03–1.26 mol% ($n = 16$), and 0.03–0.77 mol% ($n = 8$), respectively. C_1 $\delta^{13}C$ has a wide range ($-57.1‰$ to $-43.9‰$) at Joetsu Knoll (MD179-3317 and 3318C) and a narrow range ($-37.3‰$ to $-34.6‰$) at Umitaka Spur (MD179-3305G and 3306), whereas $C_1/(C_2 + C_3)$ values concentrate from 2200 to 9700 at the both areas. C_2 $\delta^{13}C$ ranged from $-31.9‰$ to $-21.3‰$ at Joetsu Knoll and from $-18.4‰$ to $-17.9‰$ at Umitaka Spur, respectively.

Table 1. Molecular compositions (mol% of the total components) and stable carbon (C_{1-3} and CO_2) and hydrogen (C_1) isotopic ratios of hydrate-bound gases retrieved from Umitaka Spur (MD179-3305G and 3306) and Joetsu Knoll (MD179-3317 and 3318C), Japan Sea. Because H_2S is highly corrosive and reaction with sampling and analytical tools was apparently not prevented, a higher portion of H_2S is lost during core retrieval and gas analysis compared to hydrocarbons and CO_2.

Core No. Depth [mbsf]	Molecular composition						Isotopic composition				
	C_1 [mol%]	C_2 [mol%]	C_3 [mol%]	CO_2 [mol%]	H_2S [mol%]	$C_1/(C_2+C_3)$	$C_1\delta^{13}C$ [‰V-PDB]	$C_2\delta^{13}C$ [‰V-PDB]	$C_3\delta^{13}C$ [‰V-PDB]	$CO_2\delta^{13}C$ [‰V-PDB]	$C_1\delta D$ [‰V-SMOW]
MD179-3305G unknown depth	98.1	0.0292	0.0019	1.10	0.77	3158	−36.6	−17.9	5.8	14.3	−167
	98.3	0.0270	0.0020	1.18	0.53	3389	−36.0	−18.2	5.7	18.0	−164
	98.2	0.0301	0.0017	1.26	0.53	3086	−36.0	−18.2	5.9	19.5	−164
	99.8	0.0331	0.0018	0.20	n.d.	2859	−34.6			17.3	−167
MD179-3306 4.5 [mbsf]	99.1	0.0292	0.0020	0.85	n.d.	3176	−37.3	−18.4	6.3	20.3	−164
	100.0	0.0221	0.0016			4213	−35.7	−18.1	4.8	23.8	−167
MD179-3317 31.5 [mbsf]	99.3	0.0171	0.0001	0.64	0.04	5783	−56.4	−31.7	−5.6	9.7	−194
	99.3	0.0153	0.0001	0.63	0.04	6446	−54.6	−31.8		11.6	−195
	99.4	0.0172	0.0001	0.57	n.d.	5758	−54.8	−31.9		8.5	−194
	99.9	0.0171	0.0001	0.11	n.d.	5810	−54.8				−194
	99.3	0.0205	0.0001	0.66	0.03	4819	−57.1	−31.4	−5.3	9.1	−194
	99.9	0.0211	0.0001	0.10	n.d.	4713	−55.4				−195
	100.0	0.0216	0.0001	0.03	n.d.	4603	−54.8				−194
MD179-3318C 1.5 [mbsf]	99.8	0.0148	0.0011	0.18	n.d.	6257	−44.0	−21.3	4.0		−189
	99.1	0.0421	0.0005	0.51	0.36	2328		−22.8	4.5	17.4	−188
	99.9	0.0456	0.0004	0.07	n.d.	2172	−44.3	−24.5			−190
	99.7	0.0078	0.0025	0.21	0.09	9682	−43.9	−21.4			n.d.

Notes: mbsf: meters below sea floor; blank: not measured; n.d.: not detected.

The molecular and stable carbon and hydrogen isotopic composition of volatiles in the headspace gas of MD179-3306 (Umitaka Spur) and 3318C (Joetsu Knoll) are summarized in Table 2. A high concentration of a compound (0.1–0.2 mol%) putatively assigned as neo-C_5 based on its relative retention time during gas chromatography was also detected. Although normal butane (n-C_4), i-C_4, isopentane (i-C_5), and normal pentane (n-C_5) were below the detection limit of the gas chromatograph, CF-IRMS (detection limit: 0.00006 mol%) detected only i-C_4 in the samples. $\delta^{13}C$ of C_5–C_8 in MD179-3306 and 3318C was within $-23‰ \pm 5‰$; however, C_3 $\delta^{13}C$ and i-C_4 $\delta^{13}C$ were exceptionally high ($+1.1‰$ to $+8.8‰$) compared with $\delta^{13}C$ of the other hydrocarbons.

Table 2. Molecular and isotopic compositions of headspace gas of the hydrate-bearing cores (MD179-3306 and MD179-3318C).

Core	MD179-3306				MD179-3318C
Depth [mbsf]	4.50	5.00	5.95	7.00	1.50
Molecular composition [mol%]					
C_1	40.7	42.3	41.4	28.4	86.9
C_2	0.0211	0.0264	0.0255	0.0153	0.0289
C_3	0.0037	0.0039	0.0037	0.0022	0.0063
neo-C_5	0.1946	0.1298	0.1862	0.0830	0.0760
CO_2	59.1	57.5	58.4	71.5	6.6
H_2S	n.d.	n.d.	n.d.	n.d.	6.3
$C_1/(C_2+C_3)$	1640	1393	1422	1624	2469
Isotopic composition [$\delta^{13}C$ ‰ V-PDB]					
C_1	−33.9	−34.6	−34.4	−34.3	−45.3
C_2	−47.3	−32.9	−21.5	−21.1	−24.8
C_3	4.5	4.5	3.9	2.9	8.8
i-C_4	2.6	4.4		2.3	1.1
neo-C_5	−23.0	−23.3	−22.8	−23.2	−21.4
2,2DMB	−18.3	−19.0	−19.3	−18.6	−15.7
2,3DMB	−21.9	−20.5		−21.6	
n-C_6	−23.0	−22.6		−23.0	
n-C_7	−28.1	−26.6		−27.4	
n-C_8	−26.4				
CO_2	31.4	31.5	31.8	31.9	−10.2
Isotopic composition [δD ‰ V-SMOW]					
C_1	−162	−166	−158	−161	−183
neo-C_5	−125	−124	−133	−125	−119

Notes: mbsf: meters below sea floor; blank: not measured; n.d.: not detected; DMB: dimethylbutane.

5. Discussion

5.1. Depth Profiles of Headspace Gas

Shallow SMI was observed in the Joetsu Basin area including Umitaka Spur and Joetsu Knoll, suggesting that the hydrocarbon flux from great depth was primarily high. The peak of the C_1 concentrations and high $C_1/(C_2 + C_3)$ values appeared around 5–8 mbsf, indicating high microbial C_1 production (methanogenesis) just beneath the SMI. Negative C_1 $\delta^{13}C$ and CO_2 $\delta^{13}C$ peaks and high H_2S

concentrations were observed around the SMI (Figure 2a–c), which typically results from the anaerobic oxidation of methane (AOM). C_1 $\delta^{13}C$ decreased around the SMI can be explained as a result of carbon recycling between the AOM and methanogenesis [33]. ANME-1 and ANME-2 groups were distinguished at Umitaka Spur and Joetsu Knoll [34] using the lipid biomarker signatures of the AOM communities [35]. In the same study area, the occurrence of AOM in the SMI zone is also supported by environmental DNA analysis [36]. Similar relations between the depth profiles of C_1 $\delta^{13}C$ and CO_2 $\delta^{13}C$ were also reported in the gassy sediment of Eckernförde Bay in the western Baltic Sea [37] and in the C_1-rich sediment near a gas chimney in northern Gulf of Mexico [38]. Although high C_1 $\delta^{13}C$ was observed in the hydrate-bearing core MD179-3318C that indicated thermogenic origin according to empirical classifications [21,39], depleted $C_1/(C_2+C_3)$, C_1 $\delta^{13}C$, and CO_2 $\delta^{13}C$ around SMI were observed even in the MD179-3318C, suggesting that C_1 is basically thermogenic and the top layer around SMI depth is affected by microbial alternation.

The upward increase in the profiles of C_1 $\delta^{13}C$ and CO_2 $\delta^{13}C$ above the SMI is due to C_1 oxidation near the sea floor [39]. On the other hand, the increase in the C_1 $\delta^{13}C$ below the SMI is explained by the concurrent increase in the CO_2 $\delta^{13}C$ with depth, resulting from the reduction of CO_2 in the methanegenic zone and the Rayleigh process. Except for MD179-3306, the profiles of C_1 δD were almost identical and gradually decreased with depth. This trend agrees fairly well with previously published δD profiles of pore water (ambient H_2O) in the same cores [40]; this is because C_1 δD is primarily determined by δD of H_2O and H_2 in the ambient water in the case of microbial C_1 generation via CO_2 reduction [41]. In contrast, the profile of C_1 δD for MD179-3306 was distinctly larger than those of other cores, indicating that C_1 is of thermogenic origin.

The concentration of C_2 in the three hydrate-bearing cores below the SMI was nearly one order of magnitude higher than that in the nonhydrate-bearing cores (Figure 2a–c), and the other cores at Umitaka Spur were also rich in C_2. These C_2-rich cores showed ^{13}C enrichment in C_2. The concentrations of C_3 in the hydrate-bearing cores and in MD179-3304 were also higher than in the nonhydrate-bearing cores. The results for C_2 and C_3 suggest two possibilities: (1) injections of thermogenic hydrocarbons from greater depth and (2) dissociation of tiny amounts of C_2- and C_3- rich sII hydrates during recovery process of sediment cores.

The concentration of C_3 in MD179-3317 was conspicuously higher (0.039–0.331 µM) than in the other cores, whereas the hydrate-bound C_3 in MD179-3317 was negligible (0.0001% of the hydrate-bound gas, Table 1). Because sI gas hydrates cannot encage C_3, it is reasonable to assume that C_3 was excluded from the hydrate formation. Similar molecular fractionations were described in previous studies [42–44].

5.2. Effect of Gas Characteristics on Hydrate Crystal

The relation between the molecular ratio of hydrate-bound hydrocarbons and the stable carbon and hydrogen isotopic ratio of individual hydrate-bound volatiles is shown in Figure 3. The Bernard plot (Figure 3a), which compares C_1 $\delta^{13}C$ with $C_1/(C_2 + C_3)$ [21], is useful for understanding the gas origin and its pathway. The data from stations MD179-3305G, 3306, and 3318C plot in the mixed gas field, whereas those from MD179-3317 plot near and within the field of microbial hydrocarbons (Figure 3a). These relation between C_1 $\delta^{13}C$ and $C_1/(C_2 + C_3)$ at Umitaka Spur and Joetsu Knoll agree well with those obtained in a previous study [22].

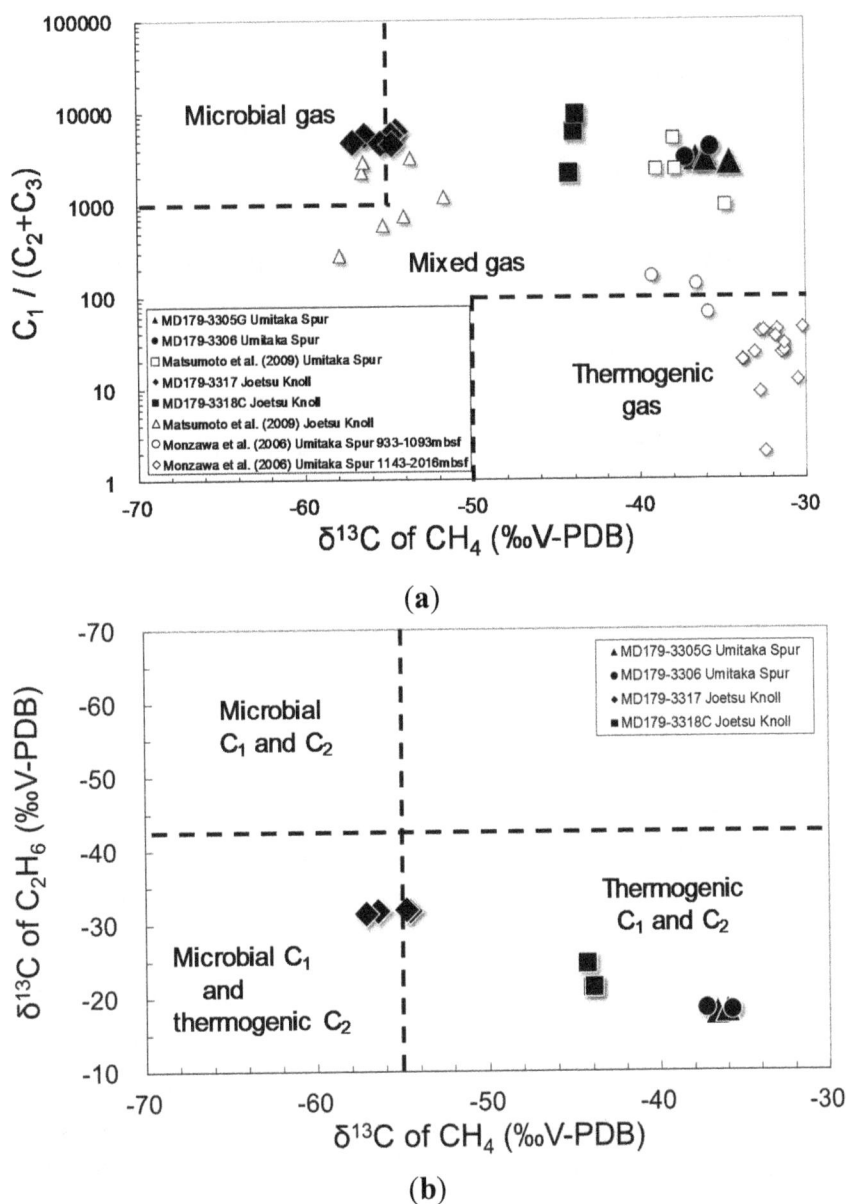

Figure 3. Relationship between molecular and isotopic composition of hydrate-bound hydrocarbons. **(a)** "Bernard plot" showing relationship between C_1 $\delta^{13}C$ and $C_1/(C_2 + C_3)$ values [21]. The data agree well with those obtained at the same area from 2005 to 2008 [22]. The headspace gas data obtained from deeper sediment layer of Umitaka Spur at depths 993–2016 m below the sea floor [20] are plotted; **(b)** Relationship between C_1 $\delta^{13}C$ and C_2 $\delta^{13}C$, based on the classification of Milkov [4].

Figure 3b shows the relation between C_1 $\delta^{13}C$ and C_2 $\delta^{13}C$. The boundaries between thermogenic and microbial origins are based on literature [4,45,46]. C_2 $\delta^{13}C$ ranged from −31.9‰ to −17.9‰, plotted in the field of thermogenic C_2 [4,45,46]. Because microbial C_2 is generally a minor component of microbial gas (<0.1%, see Figure 3a) and is depleted in ^{13}C [46], mixing of microbial and thermogenic hydrocarbons slightly decreases the C_2 $\delta^{13}C$ and strongly decreases the C_1 $\delta^{13}C$.

The molecular ratios of hydrate-bound hydrocarbons and stable carbon isotope signatures of C_1 shown in Figure 3a indicate a mixed gas origin for the hydrate at Umitaka Spur, implying depletion of C_2 and C_3 from the thermogenic gas field in Figure 3a ($C_1/(C_2 + C_3) < 100$). $C_1/(C_2 + C_3)$ values were less than 100 at depths 1143–2016 m below the sea floor at Umitaka Spur [20]. Although thermogenic gas might contain significant amounts of C_2 and C_3 (sometimes >10 mol%), as found in samples from the Gulf of Mexico [13,47], the northern Cascadia margin offshore Vancouver Island [48], and the Caspian Sea [6], the composition of C_2 and C_3 of hydrate-bound hydrocarbons at Umitaka Spur was less than 0.04% and 0.002%, respectively (Table 1). These results can be explained by molecular hydrocarbon fractionation during upward gas migration [49]. Several similar theories have been proposed, e.g., formation of C_2-rich gas hydrates in the deep layers and C_2 depletion in the residual migrating gas [22], and adsorption of C_2 and C_3 to the mineral matrix during migration [30]. Assuming steady-state upward migration of thermogenic hydrocarbons, we have to consider the consumption and/or degradation of C_2 and C_3 in the deeper sediment layers. Non-C_1 hydrocarbon degradation [43,50,51] likely selectively decreases C_2 and C_3.

Stable isotope signatures (C_1 $\delta^{13}C$, C_2 δD, and C_2 $\delta^{13}C$) of hydrate-bound and headspace gases were almost the same with each other (Figure 2a,b); however, distinct differences were found in CO_2 $\delta^{13}C$. Previous studies showed no or rather small differences in CO_2 $\delta^{13}C$ between stable isotopes of hydrate-bound and headspace gases [42,52]. It remains unsettled and further studies are needed to find the reason why such large differences in CO_2 $\delta^{13}C$ existed.

Although C_3 composition of hydrate-bound gas was very small (less than 0.0025 mol%, Table 1), we need to discuss the existence of C_3 in the hydrate-bound gas. C_1, C_2, CO_2, and H_2S can be encaged in the sI gas hydrate [1], but C_3 is exclusively found in the sII gas hydrate because of its large molecular size. It is possible that C_3 in these samples was a contaminant from the sediments attached to the hydrate during the preparation of the hydrate-bound gas. Based on the gas compositions, the gas hydrates investigated in this study were assumed to be mainly sI. This assumption is supported by the powder X-ray diffraction (PXRD) data for gas hydrates collected from Umitaka Spur and Joetsu Knoll in the past [53,54]. Nevertheless, we cannot exclude the potential coexistence of sI (C_1-rich) and sII (C_3 encaged in the large cages) gas hydrates, because PXRD cannot detect small amounts of sII. As encaged C_3 molecules in C_1-rich natural gas hydrate retrieved at the Mackenzie Delta (Onshore, Canada) were detected by Raman spectroscopy [55], heterogeneity of hydrocarbons in the different crystallographic structures is an attractive research target in the future from the viewpoint of crystal growth process of natural gas hydrates.

5.3. Neopentane and Non-Methane Hydrocarbons in the Headspace Gases

The presence of gem-dimethyl hydrocarbons was previously reported in the Deep Sea Drilling Project Holes 381 and 397 [56,57]. neo-C_5 is considered to form from the decomposition of gem-dimethylcycloalkanes derived from the terpenes of terrestrial organic matter [56]. Unusually high concentration of neo-C_5 in gases extracted from the Athabasca oil sands and proposed that neo-C_5 forms from the microbial degradation of bitumen [58]. Although the origin of neo-C_5 remains unknown, the potential involvement of microbes has been suggested [57]. Schaefer and Leythaeuser [59] explained that enrichment of neo-C_5 is caused by preferential diffusion due to the nearly spherical molecules and its diffusion coefficient, which is higher than that of less branched isomers.

Although neo-C_5 can be encaged in the large cages of sII gas hydrates [17], only trace levels of neo-C_5 were detected in the samples of hydrate-bound gas, perhaps because of contamination with the sediment gas at their retrieval process. Similar to C_3, during the formation of sI gas hydrates, *neo*-C_5 is excluded and remains in sediment. However, negligible amounts of sII gas hydrates containing *neo*-C_5 could also be present, which dissociated during core retrieval. While C_3 and *neo*-C_5 in C_1-rich hydrocarbons decrease the equilibrium pressures and stabilize their hydrate phase, the dissociation behavior of the gas hydrate depends on the guest molecules. C_1 hydrate shows self-preservation phenomena [60] in the dissociation process; however, the hydrates with lower decomposition pressures (C_2 and C_3) do not show it [61]. Further investigations are needed to understand the behavior of the small amount of sII formers, C_3 and *neo*-C_5.

C_3 $\delta^{13}C$ and *i*-C_4 $\delta^{13}C$ were exceptionally high compared with $\delta^{13}C$ of the other hydrocarbons as shown in Table 2. Possibly, ^{13}C-depleted C_3 and *i*-C_4 were preferentially consumed as microbial substrates; hence, the residual C_3 and *i*-C_4 were enriched in ^{13}C [30]. The concentration of *n*-C_4 was below the detection limit (0.00006 mol% of the original headspace sample) of the CF-IRMS; *n*-C_4 is easily transformed by hydrocarbon degraders, whereas C_2 and *i*-C_4 appear less affected or remain unaltered [42,50,51,62]. Therefore, it is reasonable to assume that *n*-C_4 was consumed primarily by biodegradation because it is more affected by biodegradation than *i*-C_4. $\delta^{13}C$ of C_1–C_8 are plotted in Figure 4: the so-called "Natural gas plot" [63]. In the Umitaka Spur area, $\delta^{13}C$ values of C_1–C_8 except C_3 and *i*-C_4 show a linear trend, indicating no microbial contribution in C_1. Moreover, the hydrate-bound C_1 comprised pure thermogenic gas.

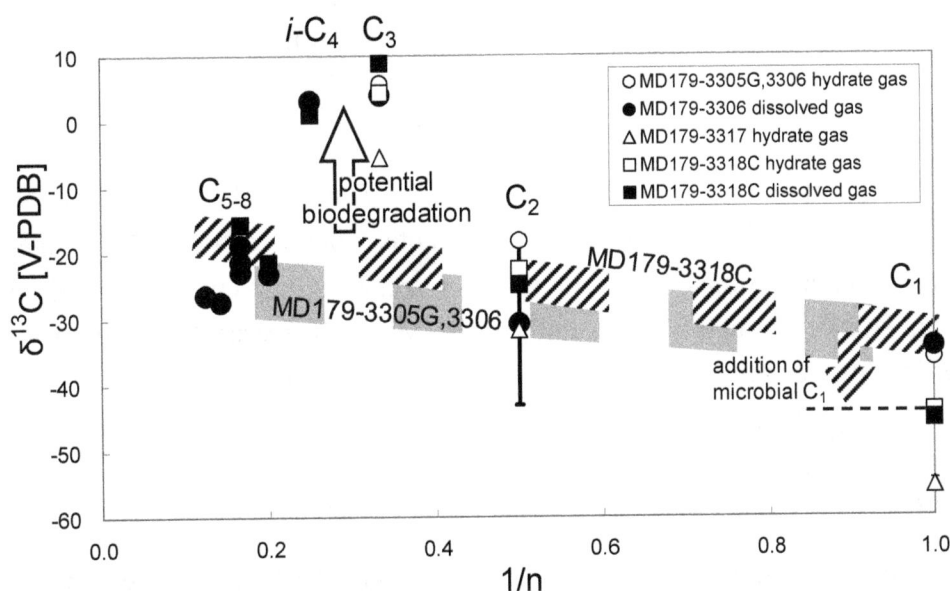

Figure 4. $\delta^{13}C$ of hydrocarbons plotted in the "Natural gas plot" adapted from [63], where n is the number of the carbon atoms of individual hydrocarbon molecules. $\delta^{13}C$ of C_1, C_2, C_3, *n*-C_4, and *n*-C_5 were plotted in the original natural gas plot. However, we plotted the data of *i*-C_4 (instead of *n*-C_4), *neo*-C_5 (instead of *n*-C_5), and C_{6-8} shown in Table 2. The concentrations of *n*-C_4, and *n*-C_5 were below the detection limit. Thick shaded lines show linear relation between C_2 and C_8 for MD179-3305G and 3306 (Umitaka Spur) and for MD179-3318C (Joetsu Knoll), except ^{13}C-rich C_3 and *i*-C_4 due to potential biodegradation. C_2 $\delta^{13}C$ of MD179-3306 has a large error bar because of wide distribution along with depth (Table 2).

In contrast, these gases mixed with microbial C_1 during migration to shallow sediment layers in the Joetsu Knoll area, as C_1 $\delta^{13}C$ decreased from the anticipated thermogenic C_1 value for MD179-3318C.

6. Conclusions

Molecular and stable carbon and hydrogen isotopic compositions of hydrate-bound and pore-water gas were reported for samples retrieved from Umitaka Spur and Joetsu Knoll at the eastern margin of the Japan Sea. According to empirical classifications of gas data, the hydrate-bound gas from Umitaka Spur was of thermogenic origin and that from Joetsu Knoll partly contained microbial gas. C_1 concentration depth profiles showed shallow SMI in these areas, indicating high C_1 flux. The molecular composition of C_2 and C_3 for the headspace gas of hydrate-bound cores were around a thousandth smaller than those of hydrate-bound thermogenic gases across the world, and C_3 $\delta^{13}C$ and i-C_4 $\delta^{13}C$ were exceptionally high (+1.1‰ to +8.8‰) in the pore water, suggesting that biodegradation affects these hydrocarbons in hydrate-bearing sediment systems.

High concentration of neo-C_5 was detected in the pore water gases in MD179-3306 (Umitaka Spur) and 3318C (Joetsu Knoll), which suggests that neo-C_5 was excluded from the hydrate crystal during the formation process and thus remained in the pore water, because neo-C_5 cannot be encaged by sI gas hydrates. This is supported by the observation that the concentration of C_3 in the hydrate-bearing core was higher than that in the nonhydrate-bearing cores. However, small amounts of sII gas hydrate can possibly encage these heavier hydrocarbons. Because neo-C_5 forms from the microbial degradation of organic matter and is easy to diffuse in sediment layers, the accumulation of neo-C_5 in subsurface sediment can affect and change the crystallographic properties of gas hydrates in marine sediments.

Acknowledgments

We appreciate the support of the crew onboard R/V *Marion Dufresne* during the MD179/Japan Sea Gas Hydrate cruise. We express our gratitude to the scientists onboard the MD179/Japan Sea Gas Hydrate cruise. We also thank Thomas Pape (University of Bremen, Germany) for valuable suggestions and comments. This study was mainly supported by the MH21 Research Consortium for Methane Hydrate Resources in Japan, and the analytical system was supported by the Grant-in-Aid for Scientific Research (C) 22540485 and (B) 26303021 of the Japan Society for the Promotion of Science (JSPS).

Author Contributions

Akihiro Hachikubo designed the study, performed gas analysis, and drafted the manuscript. Katsunori Yanagawa (microbiology), Hitoshi Tomaru (pore water geochemistry), and Hailong Lu (crystallography) helped to draft the manuscript from their professional point of view. Ryo Matsumoto designed the framework of the MD179 cruise. All authors discussed the results and approved the final manuscript.

Conflicts of Interest

The authors declare no conflict of interest.

References

1. Sloan, E.D.; Koh, C.A. *Clathrate Hydrates of Natural Gases*, 3rd ed.; CRC Press: Boca Raton, FL, USA, 2008.

2. Kvenvolden, K.A. Potential effects of gas hydrate on human welfare. *Proc. Natl. Acad. Sci. USA* **1999**, *96*, 3420–3426.

3. Judd, A.G.; Hovland, M.; Dimitrov, L.I.; García Gil, S.; Jukes, V. The geological methane budget at continental margins and its influence on climate change. *Geofluids* **2002**, *2*, 109–126.

4. Milkov, A.V. Molecular and stable isotope compositions of natural gas hydrates: A revised global dataset and basic interpretations in the context of geological settings. *Org. Geochem.* **2005**, *36*, 681–702.

5. Paull, C.K.; Ussler, W., III; Borowski, W.S.; Spiess, F.N. Methane-rich plumes on the Carolina continental rise: Associations with gas hydrates. *Geology* **1995**, *23*, 89–92.

6. Ginsburg, G.D.; Soloviev, V.A. *Submarine Gas Hydrates*; VNIIOkeangeologia: St. Petersburg, Russia, 1998.

7. Heeschen, K.U.; Tréhu, A.M.; Collier, R.W.; Suess, E.; Rehder, G. Distribution and height of methane bubble plumes on the Cascadia Margin characterized by acoustic imaging. *Geophys. Res. Lett.* **2003**, *30*, doi:10.1029/2003GL016974.

8. Kvenvolden, K.A. Gas hydrates—Geological perspective and global change. *Rev. Geophys.* **1993**, *31*, 173–187.

9. Sloan, E.D., Jr. Fundamental principles and applications of natural gas hydrates. *Nature* **2003**, *426*, 353–363.

10. Boswell, R.; Collett, T.S. Current perspectives on gas hydrate resources. *Energy Environ. Sci.* **2011**, *4*, 1206–1215.

11. Subramanian, S.; Kini, R.A.; Dec, S.F.; Sloan, E.D. Evidence of structure II hydrate formation from methane + ethane mixtures. *Chem. Eng. Sci.* **2000**, *55*, 1981–1999.

12. Subramanian, S.; Ballard, A.L.; Kini, R.A.; Dec, S.F.; Sloan, E.D. Structural transitions in methane + ethane gas hydrates—Part I: Upper transition point and applications. *Chem. Eng. Sci.* **2000**, *55*, 5763–5771.

13. Brooks, J.M.; Kennicutt, M.C., II; Fay, R.R.; McDonald, T.J.; Sassen, R. Thermogenic gas hydrates in the Gulf of Mexico. *Science* **1984**, *225*, 409–411.

14. Bourry, C.; Chazallon, B.; Charlou, J.L.; Donval, J.P.; Ruffine, L.; Henry, P.; Geli, L.; Çagatay, M.N.; İnan, S.; Moreau, M. Free gas and gas hydrates from the Sea of Marmara, Turkey. Chemical and structural characterization. *Chem. Geol.* **2009**, *264*, 197–206.

15. Kida, M.; Khlystov, O.; Zemskaya, T.; Takahashi, N.; Minami, H.; Sakagami, H.; Krylov, A.; Hachikubo, A.; Yamashita, S.; Shoji, H.; *et al.* Coexistence of structure I and II gas hydrates in Lake Baikal suggesting gas sources from microbial and thermogenic origin. *Geophys. Res. Lett.* **2006**, *33*, doi:10.1029/2006GL028296.

16. Kida, M.; Hachikubo, A.; Sakagami, H.; Minami, H.; Krylov, A.; Yamashita, S.; Takahashi, N.; Shoji, H.; Khlystov, O.; Poort, J.; *et al.* Natural gas hydrates with locally different cage occupancies and hydration numbers in Lake Baikal. *Geochem. Geophys. Geosyst.* **2009**, *10*, doi:10.1029/2009GC002473.

17. Davidson, D.W.; Garg, S.K.; Gough, S.R.; Hawkins, R.E.; Ripmeester, J.A. Characterization of natural gas hydrates by nuclear magnetic resonance and dielectric relaxation. *Can. J. Chem.* **1977**, *55*, 3641–3650.

18. Aoyama, C.; Matsumoto, R. Acoustic surveys of methane plumes by quantitative echo sounder in Japan Sea and the estimate of the seeping amount of the methane hydrate bubbles. *J. Geogr.* **2009**, *118*, 156–174. (In Japanese)

19. Matsumoto, R.; Okuda, Y.; Aoyama, C.; Hiruta, A.; Ishida, Y.; Sunamura, M.; Numanami, H.; Tomaru, H.; Snyder, G.; Komatsubara, J.; *et al.* Methane plumes over a marine gas hydrate system in the eastern margin of Japan Sea: A possible mechanism for the transportation of subsurface methane to shallow waters. In Proceedings of the 5th International Conference on Gas Hydrates, Trondheim, Norway, 13–16 June 2005; Volume 3006, pp. 749–754.

20. Monzawa, N.; Kaneko, M.; Osawa, M. A review of petroleum system in the deep water area of the Toyama Trough to the Sado Island in the Japan Sea, based on the results of the METI Sado Nansei Oki drilling. *J. Jpn. Assoc. Pet. Technol.* **2006**, *71*, 618–627. (In Japanese)

21. Bernard, B.B.; Brooks, J.M.; Sackett, W.M. Natural gas seepage in the Gulf of Mexico. *Earth Planet. Sci. Lett.* **1976**, *31*, 48–54.

22. Matsumoto, R.; Okuda, Y.; Hiruta, A.; Tomaru, H.; Takeuchi, E.; Sanno, R.; Suzuki, M.; Tsuchinaga, K.; Ishida, Y.; Ishizaki, O.; *et al.* Formation and collapse of gas hydrate deposits in high methane flux area of the Joetsu Basin, eastern margin of Japan Sea. *J. Geogr.* **2009**, *118*, 43–71. (In Japanese)

23. Hachikubo, A.; Yanagawa, K.; Tomaru, H.; Matsumoto, R. Dissolved gas analysis of pore water in subsurface sediments retrieved at eastern margin of Japan Sea (MD179 gas hydrates cruise). *J. Jpn. Assoc. Pet. Technol.* **2012**, *77*, 268–273. (In Japanese)

24. Saeki, T.; Inamori, T.; Nagakubo, S.; Ward, P.; Asakawa, E. 3D seismic velocity structure below mounds and pockmarks in the deep water southwest of the Sado Island. *J. Geogr.* **2009**, *118*, 93–110. (In Japanese)

25. Hiruta, A.; Snyder, G.T.; Tomaru, H.; Matsumoto, R. Geochemical constraints for the formation and dissociation of gas hydrate in an area of high methane flux, eastern margin of the Japan Sea. *Earth Planet. Sci. Lett.* **2009**, *279*, 326–339.

26. Machiyama, H.; Kinoshita, M.; Takeuchi, R.; Matsumoto, R.; Yamano, M.; Hamamoto, H.; Hiromatsu, M.; Satoh, M.; Komatsubara, J. Heat flow distribution around the Joetsu gas hydrate field, western Joetsu basin, eastern margin of the Japan Sea. *J. Geogr.* **2009**, *118*, 986–1007. (In Japanese)

27. Freire, A.F.M.; Menezes, T.R.; Matsumoto, R.; Sugai, T.; Miller, D.J. Origin of the organic matter in the Late Quaternary sediments of the eastern margin of Japan Sea. *J. Sedimentol. Soc. Jpn.* **2009**, *68*, 117–128.

28. Hachikubo, A.; Krylov, A.; Sakagami, H.; Minami, H.; Nunokawa, Y.; Shoji, H.; Matveeva, T.; Jin, Y.K.; Obzhirov, A. Isotopic composition of gas hydrates in subsurface sediments from offshore Sakhalin Island, Sea of Okhotsk. *Geo-Mar. Lett.* **2010**, *30*, 313–319.

29. Hachikubo, A.; Khlystov, O.; Krylov, A.; Sakagami, H.; Minami, H.; Nunokawa, Y.; Yamashita, S.; Takahashi, N.; Shoji, H.; Nishio, S.; *et al.* Molecular and isotopic characteristics of gas hydrate-bound hydrocarbons in southern and central Lake Baikal. *Geo-Mar. Lett.* **2010**, *30*, 321–329.

30. Waseda, A.; Iwano, H. Reservoir evaluation using carbon isotope composition of gas. *J. Jpn. Assoc. Pet. Technol.* **2007**, *72*, 585–593. (In Japanese)

31. Sakagami, H.; Takahashi, N.; Hachikubo, A.; Minami, H.; Yamashita, S.; Shoji, H.; Khlystov, O.; Kalmychkov, G.; Grachev, M.; de Batist, M. Molecular and isotopic composition of hydrate-bound and sediment gases in the southern basin of Lake Baikal, based on an improved headspace gas method. *Geo-Mar. Lett.* **2012**, *32*, 465–472.

32. Gealy, E.L.; Dubois, R. Shipboard geochemical analysis, Leg 7, Glomar Challenger. *Initial Rep. Deep Sea Drill. Proj.* **1971**, *7*, 863–869.

33. Knab, N.J.; Cragg, B.A.; Hornibrook, E.R.C.; Holmkvist, L.; Pancost, R.D.; Borowski, C.; Parkes, R.J.; Jørgensen, B.B. Regulation of anaerobic methane oxidation in sediments of the Black Sea. *Biogeosciences* **2009**, *6*, 1505–1518.

34. Ogihara, S.; Ishizaki, O.; Matsumoto, R. Organic geochemical analysis of push core sediment samples collected from NT-06-19 (Umitaka Spur and Joetsu Knoll off Naoetsu). *J. Geogr.* **2009**, *118*, 128–135. (In Japanese)

35. Niemann, H.; Elvert, M. Diagnostic lipid biomarker and stable carbon isotope signatures of microbial communities mediating the anaerobic oxidation of methane with sulphate. *Org. Geochem.* **2008**, *39*, 1668–1677.

36. Yanagawa, K.; Sunamura, M.; Lever, M.A.; Morono, Y.; Hiruta, A.; Ishizaki, O.; Matsumoto, R.; Urabe, T.; Inagaki, F. Niche separation of methanotrophic archaea (ANME-1 and -2) in methane-seep sediments of the Eastern Japan Sea offshore Joetsu. *Geomicrobiol. J.* **2011**, *28*, 118–129.

37. Martens, C.S.; Albert, D.B.; Alperin, M.J. Stable isotope tracing of anaerobic methane oxidation in the gassy sediments of Eckernförde Bay, German Baltic Sea. *Am. J. Sci.* **1999**, *299*, 589–610.

38. Ussler, W., III; Paull, C.K. Rates of anaerobic oxidation of methane and authigenic carbonate mineralization in methane-rich deep-sea sediments inferred from models and geochemical profiles. *Earth Planet. Sci. Lett.* **2008**, *266*, 271–287.

39. Whiticar, M.J. Carbon and hydrogen isotope systematics of bacterial formation and oxidation of methane. *Chem. Geol.* **1999**, *161*, 291–314.

40. Tomaru, H.; Hachikubo, A.; Yanagawa, K.; Muramatsu, Y.; Anzai, H.; Snyder, G.T.; Matsumoto, R. Geochemistry of pore waters from gas hydrate research in the eastern margin of the Japan Sea (MD179). *J. Jpn. Assoc. Pet. Technol.* **2012**, *77*, 262–267. (In Japanese)

41. Kawagucci, S.; Kobayashi, M.; Hattori, S.; Yamada, K.; Ueno, Y.; Takai, K.; Yoshida, N. Hydrogen isotope systematics among H_2–H_2O–CH_4 during the growth of the hydrogenotrophic methanogen *Methanothermobacter thermautotrophicus* strain ΔH. *Geochim. Cosmochim. Acta* **2014**, *142*, 601–614.

42. Pape, T.; Bahr, A.; Rethemeyer, J.; Kessler, J.D.; Sahling, H.; Hinrichs, K.-U.; Klapp, S.A.; Reeburgh, W.S.; Bohrmann, G. Molecular and isotopic partitioning of low-molecular-weight hydrocarbons during migration and gas hydrate precipitation in deposits of a high-flux seepage site. *Chem. Geol.* **2010**, *269*, 350–363.

43. Milkov, A.V.; Claypool, G.E.; Lee, Y.-J.; Torres, M.E.; Borowski, W.S.; Tomaru, H.; Sassen, R.; Long, P.E. ODP Leg 204 Scientific Party Ethane enrichment and propane depletion in subsurface gases indicate gas hydrate occurrence in marine sediments at southern Hydrate Ridge offshore Oregon. *Org. Geochem.* **2004**, *35*, 1067–1080.

44. Sassen, R.; Sweet, S.T.; DeFreitas, D.A.; Milkov, A.V. Exclusion of 2-methylbutane (isopentane) during crystallization of structure II gas hydrate in sea-floor sediment, Gulf of Mexico. *Org. Geochem.* **2000**, *31*, 1257–1262.

45. Sassen, R.; Curiale, J.A. Microbial methane and ethane from gas hydrate nodules of the Makassar Strait, Indonesia. *Org. Geochem.* **2006**, *37*, 977–980.

46. Taylor, S.W.; Sherwood Lollar, B.; Wassenaar, L.I. Bacteriogenic ethane in near-surface aquifers: Implications for leaking hydrocarbon well bores. *Environ. Sci. Technol.* **2000**, *34*, 4727–4732.

47. Sassen, R.; Joye, S.; Sweet, S.T.; DeFreitas, D.A.; Milkov, A.V.; MacDonald, I.R. Thermogenic gas hydrates and hydrocarbon gases in complex chemosynthetic communities, Gulf of Mexico continental slope. *Org. Geochem.* **1999**, *30*, 485–497.

48. Pohlman, J.W.; Canuel, E.A.; Chapman, N.R.; Spence, G.D.; Whiticar, M.J.; Coffin, R.B. The origin of thermogenic gas hydrates on the northern Cascadia Margin as inferred from isotopic (13C/12C and D/H) and molecular composition of hydrate and vent gas. *Org. Geochem.* **2005**, *36*, 703–716.

49. Lorenson, T.D.; Whiticar, M.J.; Waseda, A.; Dallimore, S.R.; Collett, T.S. Gas composition and isotopic geochemistry of cuttings, core, and gas hydrate from the JAPEX/JNOC/GSC Mallik 2L-38 gas hydrate research well. *Geol. Surv. Can. Bull.* **1999**, *544*, 143–163.

50. James, A.T.; Burns, B.J. Microbial alteration of subsurface natural gas accumulations. *Am. Assoc. Pet. Geol. Bull.* **1984**, *68*, 957–960.

51. Kniemeyer, O.; Musat, F.; Sievert, S.M.; Knittel, K.; Wilkes, H.; Blumenberg, M.; Michaelis, W.; Classen, A.; Bolm, C.; Joye, S.B.; *et al.* Anaerobic oxidation of short-chain hydrocarbons by marine sulphate-reducing bacteria. *Nature* **2007**, *449*, 898–901.

52. Kim, J.-H.; Torres, M.E.; Choi, J.; Bahk, J.-J.; Park, M.-H.; Hong, W.-L. Influences on gas transport based on molecular and isotopic signatures of gases at acoustic chimneys and background sites in the Ulleung Basin. *Org. Geochem.* **2012**, *43*, 26–38.

53. Lu, H.; Moudrakovski, I.L.; Matsumoto, R.; Dutrisac, R.; Ripmeester, J.A. The characteristics of gas hydrates recovered from shallow sediments at Umitaka spur, Eastern margin of the Sea of Japan. In Proceedings of the American Geophysical Union Fall Meeting 2008, San Francisco, CA, USA, 15–19 December 2008.

54. Lu, H.; Moudrakovski, I.L.; Ripmeester, J.A.; Ratcliffe, C.I.; Matsumoto, R.; Tani, A. The characteristics of gas hydrates recovered from Joetsu basin, eastern margin of the Sea of Japan. In Proceedings of the 7th International Conference on Gas Hydrates, Edinburgh, UK, 17–21 July 2011.

55. Uchida, T.; Uchida, T.; Kato, A.; Sasaki, H.; Kono, F.; Takeya, S. Physical properties of natural gas hydrate and associated gas-hydrate-bearing sediments in the JAPEX/JNOC/GSC *et al.* Mallik 5L-38 gas hydrate production research well. *Geol. Surv. Can. Bull.* **2005**, *585*, 1–10.

56. Hunt, J.M.; Whelan, J.K. Dissolved gases in Black Sea sediments. *DSDP Initial Rep.* **1978**, *42*, 661–665.

57. Whelan, J.K. C$_1$ to C$_7$ hydrocarbons from IPOD holes 397 and 397A. *DSDP Initial Rep.* **1979**, *47*, 531–539.

58. Strausz, O.P.; Jha, K.N.; Montgomery, D.S. Chemical composition of gases in Athabasca bitumen and in low-temperature thermolysis of oil sand, asphaltene and maltene. *Fuel* **1977**, *56*, 114–120.

59. Schaefer, R.G.; Leythaeuser, D. C$_2$–C$_8$ hydrocarbons in sediments from Deep Sea Drilling Project Leg 75, holes 530A, Angola Basin, and 532, Walvis Ridge. *DSDP Initial Rep.* **1984**, *75*, 1055–1067.

60. Yakushev, V.S.; Istomin, V.A. Gas-hydrates self-preservation effect. In *Physics and Chemistry of Ice*; Hokkaido University Press: Sapporo, Japan, 1992; pp. 136–140.

61. Takeya, S.; Ripmeester, J.A. Dissociation behavior of clathrate hydrates to ice and dependence on guest molecules. *Angew. Chem. Int. Ed.* **2008**, *47*, 1276–1279.

62. Wang, W.-C.; Zhang, L.-Y.; Liu, W.-H.; Kang, Y.; Ren, J.-H. Effects of biodegradation on the carbon isotopic composition of natural gas—A case study in the Bamianhe oil field of the Jiyang Depression, Eastern China. *Geochem. J.* **2005**, *39*, 301–309.

63. Chung, H.M.; Gormly, J.R.; Squires, R.M. Origin of gaseous hydrocarbons in subsurface environments: Theoretical considerations of carbon isotope distribution. *Chem. Geol.* **1988**, *71*, 97–103.

7

Wheel Torque Distribution of Four-Wheel-Drive Electric Vehicles Based on Multi-Objective Optimization

Cheng Lin [†] and Zhifeng Xu [†,*]

Collaborative Innovation Center of Electric Vehicles in Beijing, Beijing Institute of Technology, Beijing 100081, China; E-Mail: lincheng@bit.edu.cn

[†] These authors contributed equally to this work.

[*] Author to whom correspondence should be addressed; E-Mail: xzf2012@126.com

Academic Editor: Joeri Van Mierlo

Abstract: The wheel driving torque on four-wheel-drive electric vehicles (4WDEVs) can be modulated precisely and continuously, therefore maneuverability and energy-saving control can be carried out at the same time. In this paper, a wheel torque distribution strategy is developed based on multi-objective optimization to improve vehicle maneuverability and reduce energy consumption. In the high-layer of the presented method, sliding mode control is used to calculate the desired yaw moment due to the model inaccuracy and parameter error. In the low-layer, mathematical programming with the penalty function consisting of the yaw moment control offset, the drive system energy loss and the slip ratio constraint is used for wheel torque control allocation. The programming is solved with the combination of off-line and on-line optimization to reduce the calculation cost, and the optimization results are sent to motor controllers as torque commands. Co-simulation based on MATLAB® and Carsim® proves that the developed strategy can both improve the vehicle maneuverability and reduce energy consumption.

Keywords: 4WDEV; wheel torque distribution; control allocation

1. Introduction

As automobile emission standards are getting more stringent [1], research on and the application of electric drive technology has become a hot topic. Electric drive systems are zero emission and their output torque can be modulated precisely [2]. Based on the structure of drive system, electric vehicles can be categorized into centrally driven and distributed driven ones. The continuous and precise modulation of each wheel driving torque in the 4WDEV gives it an advantage over centrally driven vehicles when carrying out vehicle motion control and in energy savings. For example, the vehicle yaw rate can be modified through differential control of wheels on distributed driven vehicles [3], moreover, properly distributing wheel torque under different circumstances can achieve high energy efficiency which meets the energy saving needs.

The drive system structure of a four-wheel-drive electric vehicle [4] is shown in Figure 1. In this paper, a dedicated torque-vectoring controller that allows optimal wheel torque allocation is developed based on a hierarchical structure. In the high-layer, the desired driving torque and yaw moment are figured out and sent to the low-layer. In the low-layer, wheel torque is distributed for both improving the vehicle maneuverability and reducing the energy consumption.

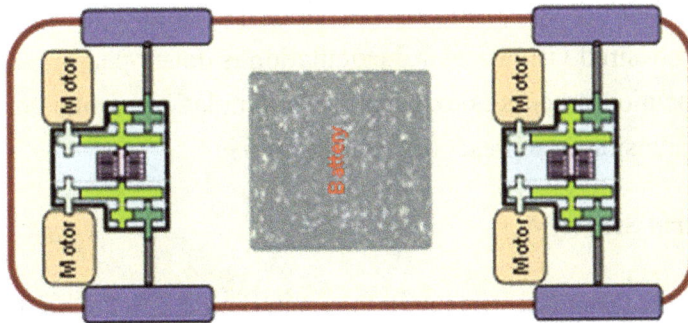

Figure 1. Drive system structure of the nominal vehicle.

As shown in the literature, many methods have been used to calculate the desired yaw moment in vehicle yaw motion control including sliding mode control [5] (SMC), fuzzy logic control [6], feed forward and feedback control [7], H_∞ robust control [8], model predictive control [9] and adaptive control [10]. In this paper, SMC is adopted in the high-level controller due to its greater robustness to model inaccuracy and parameter error.

In terms of the wheel torque distribution layer, typical methods introduced in the literature are as follows: a torque distribution method was proposed in [11] to show the potential of optimizing the 4WDEV operational energy efficiency. The authors deduced the driving torque summation of the two wheels on the left or right side of the vehicle according to the desired yaw moment and the total driving torque, given the angle of steering wheels were approximately zero. Simulation results verified that the proposed strategy could improve the energy efficiency of the drive system, but there was no description of the influence of control strategy on vehicle maneuverability, and no slip ratio constraint was considered which may cause excessive spin of the driving wheels. In [12,13], the authors developed a wheel torque control strategy based on an optimization algorithm, and adaptive energy efficient control allocation (A-EECA) was adopted to reduce the calculation cost. The authors pointed out that the wheel torque distribution was not necessarily optimized at each time step and A-EECA can distribute wheel torque trending in the optimal direction at each step. As the optimization at each step was started from the last torque state, the action of torque distribution cannot follow virtual control changes in a timely way, as shown in Figure 6 of [12] and Figure 6, Figure 7 of [13]. Moreover, the influence of the controller on vehicle maneuverability was not shown clearly in the papers above. In [14], the authors pointed out that the optimal torque distribution must rely on the appropriate selection of the penalty function, and the vehicle performance provided by alternative penalty functions for the optimal wheel torque distribution of a 4WDEV were evaluated based on an off-line optimization algorithm. Results showed that using penalty functions based on the minimum tire slip criterion can achieve better vehicle performance than using functions based on energy efficiency, but the authors did not elaborate the practical algorithm for on-line application.

In this paper, the wheel torque is distributed based on multi-objective optimization. Mathematical programming is adopted with the penalty function consisting of yaw moment control offset, drive system energy loss and slip ratio constraint. The Newton-Lagrange algorithm is used to search for the optimal point on-line starting from the off-line optimization result. The on-line results are sent to the motor controllers as wheel torque commands. Co-simulation based on MATLAB® and Carsim® verifies the proposed strategy in terms of vehicle motion control and energy saving. The rest of this paper is organized as follows: in Section 2, the proposed wheel torque control strategy is described. The high-level controller used for desired virtual control calculation is developed in Section 3. In Section 4, the wheel torque control allocation process is explained. Co-simulation results are shown in Section 5 followed by discussions. Conclusions are presented in Section 6.

2. The Wheel Torque Control Strategy

The work flow diagram of the proposed strategy is shown in Figure 2. In the high-layer of the strategy, the desired driving torque and yaw moment are figured out. The driver intention is interpreted first, and the desired yaw rate is obtained.

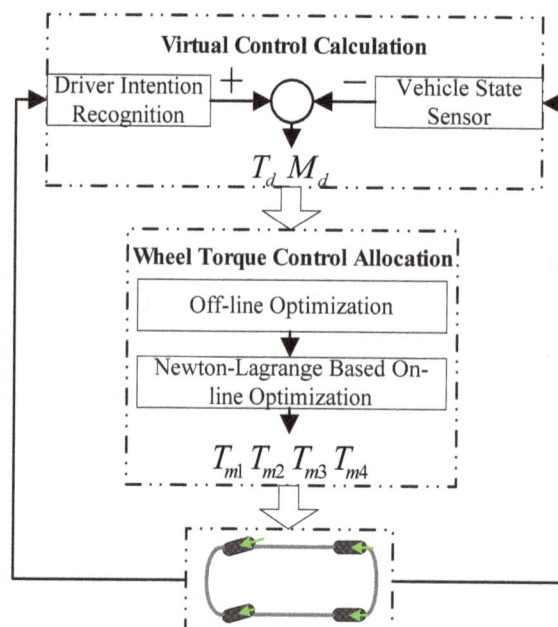

Figure 2. The wheel torque control strategy for 4WDEV.

After that the difference between the desired yaw rate and vehicle real yaw rate observed by sensors is obtained to calculate the desired yaw moment based on SMC. In the lower-layer, the desired yaw moment and driving torque will be used in a multi-objective mathematical programming, where the penalty function consists of the yaw moment control offset, drive system energy loss and wheel slip ratio constraint. During wheel torque control allocation the maneuverability improvement and energy-saving control are both taken into consideration and the four wheels' slip ratios are balanced. As the penalty function is nonlinear and non-convex, the calculation cost of algorithms to figure out the global optimal point is too high for on-line application. In this paper, off-line optimization based on a simplified penalty function is used to preliminarily get an optimal point, and the following on-line optimization is used to search for the local optimal point starting from the off-line result. The results of the on-line optimization will be used as wheel torque commands.

3. Desired Yaw Moment Calculation

The wheel torque distribution strategy presented in this paper works while the vehicle is accelerating and cornering. Firstly, a control-oriented vehicle model is built. Based on Figure 3, the yaw motion of the vehicle can be written as:

$$M = I_z \dot{\Omega} \tag{1}$$

where M is the generalized external moment acting about the Z-axis, I_z is vehicle moment of inertia about the Z-axis, Ω is the yaw rate of the vehicle.

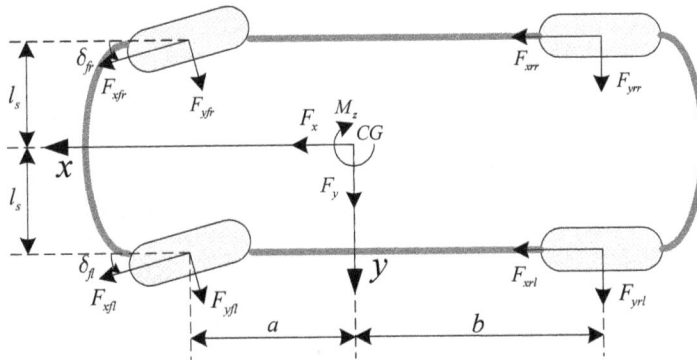

Figure 3. Coordinates for planar motions of the 4WDEV.

On the 4WDEV, the accelerator pedal input and steering wheel input can be used to recognize driver intention. In the high-layer of the control strategy, the desired torque is figured out from driver input. A model suggested in [15] is used as reference model of desired yaw rate as follows:

$$\frac{\Omega_d(s)}{\Theta_{sw}(s)} = \frac{\widehat{V}_x}{(\tau_r s + 1)(\tau_d s + 1)} \tag{2}$$

where $\widehat{V}_x = \frac{k_r V_x}{(1 + k_a V_x^2) l} GR$. Ω_d is the desired vehicle yaw rate, Θ_{sw} is the steering wheel input, t_r and τ_d are the time constants, V_x is the vehicle velocity along the X-axis, k_r is the gain of the reference model, k_a is the stability factor, l is the vehicle wheelbase, GR is the gear ratio of the front steering mechanism linkage.

The difference between the desired yaw rate and observed value by sensors is used to calculate the desired yaw moment based on SMC as follows. In the SMC framework adopted in this work, the control objective is to reach and remain in sliding surface $e = 0$, where:

$$e = \Omega - \Omega_d + \int (\Omega - \Omega_d) dt \tag{3}$$

First-order asymptotically stable desired error dynamics is defined as follows:

$$\dot{e} = -k \, \mathrm{sgn}(e) \tag{4}$$

where k is the gain factor.

Then the desired yaw moment M_d is:

$$M_d = I_z \left[\dot{\Omega}_d - (\Omega - \Omega_d) - k \, \mathrm{sgn}(e) \right] \tag{5}$$

and:

$$
\begin{aligned}
M = & F_{xfl}(a\sin\delta_{fl} - l_s\cos\delta_{fl}) + F_{yfl}(a\cos\delta_{fl} + l_s\sin\delta_{fl}) + F_{xfr}(a\sin\delta_{fr} + l_s\cos\delta_{fr}) \\
& + F_{yfr}(a\cos\delta_{fr} - l_s\sin\delta_{fr}) - F_{xrl}l_s + F_{yrl}b + F_{xrr}l_s - F_{yrr}b
\end{aligned}
\tag{6}
$$

where F_x is the tire longitudinal force, F_y is the tire lateral force. a, b is the distance from vehicle center of gravity (c.g.) to the front and rear axle, respectively. δ_{fl} and δ_{fr} are the steering angle of the front left and front right wheel, respectively. The track is shown in Figure 3 as $2l_s$. The corner marks fl, fr, rl, rr mean the front left, front right, rear left and rear right wheel, respectively.

The simplified equation of motion for wheel is:

$$T_{mi}i_r - F_{xi}R_r = J_\omega \dot{\omega}_i \tag{7}$$

where T_{mi} is the motor torque, i_r is the gear ratio, R_r is the wheel radius, J_ω is the wheel rotational inertia, ω_i is wheel angular velocity and $i = 1, 2, 3$ or 4 corresponds to the fl, fr, rl or rr wheel, respectively.

When $\lambda \ll 1$, $J\dot{\omega}_i$ is sufficiently small compared to $T_{mi}i_r$ and $F_{xi}R_r$, therefore:

$$T_{mi}i_r \approx F_{xi}R_r \tag{8}$$

Then we can get:

$$
\begin{aligned}
& T_{mfl}(a\sin\delta_{fl} - L_s\cos\delta_{fl}) + T_{mfr}(L_s\cos\delta_{fr} + a\sin\delta_{fr}) + T_{mrl}(-L_s) + T_{mrr}L_s \\
& = M_d R_r - F_{yfl}R_r(a\cos\delta_F + L_s\sin\delta_F) - F_{yfr}R_r(a\cos\delta_R - L_s\sin\delta_R) + F_{yrl}bR_r + F_{yrr}bR_r
\end{aligned}
\tag{9}
$$

As the developed controller works when the vehicle is driven in the linear region with small tire slip angle, the following approximation is used here:

$$F_{yi} = -C_i\alpha_i \tag{10}$$

where C_i is the wheel cornering stiffness, α_i is the tire slip angle. C_i and α_i can be obtained by a sensor or observer. Thus we can get an equation which defines the ideal relation between wheel driving torque T_{mi} and desired vehicle yaw rate Ω_d:

$$\sum_{i=1}^{4} T_{mi} L_i = I_z R_r \left[\dot{\Omega}_d - (\Omega - \Omega_d) - k \operatorname{sgn}(e) \right]$$

$$-F_{yfl} R_r (a \cos \delta_{fl} + L_s \sin \delta_{fl}) - F_{yfr} R_r (a \cos \delta_{fr} - L_s \sin \delta_{fr}) + F_{yrl} b R_r + F_{yrr} b R_r$$ (11)

In addition, we can get the total driving torque T_d requested by the driver from:

$$T_d = 4 P T_{m\max}$$ (12)

where P is the throttle opening, $T_{m\max}$ is the motor peak torque. From Equations (11) and (12) we can get the constraints for wheel driving torque. It is important to note here that Equation (11) does not have to be met.

4. Wheel Torque Control Allocation

4.1. The Multi-Objective Optimization Algorithm

In the low-layer of the proposed wheel torque control strategy, a multi-objective mathematical programming method is adopted. The penalty function consisting of yaw moment control offset, drive system energy loss and slip ratio constraint is shown as follows:

$$\min_{T_{mi}} J = (\sum_{i=1}^{4} T_{mi} L_i - V_d)^2 + \sigma_p \sum_{i=1}^{4} C_p(T_{mi}) + \sigma_t C_t(T_{mi})$$

$$s.t. \ \sum_{i=1}^{4} T_{mi} = T_d;$$ (13)

$$T_{mi\min} \le T_{mi} \le T_{mi\max}, \ i = 1, 2, 3, 4.$$

The first part of the penalty function J is yaw moment control offset:

$$\Delta_M = (\sum_{i=1}^{4} T_{mi} L_i - V_d)^2$$ (14)

where: $L_1 = a \sin \delta_F - L_s \cos \delta_F$, $L_2 = L_s \cos \delta_R + a \sin \delta_R$, $L_3 = -L_s$, $L_4 = L_s$,
$V_d = M_d R_r - F_{y1} R_r (a \cos \delta_F + L_s \sin \delta_F) - F_{y2} R_r (a \cos \delta_R - L_s \sin \delta_R) + F_{y3} b R_r + F_{y4} b R_r$.

This is the difference between the yaw moment generated by the current torque distribution point and the desired yaw moment from high-layer yaw motion controller. This is used to control the vehicle yaw rate, which will optimize the vehicle cornering response.

The second part is the energy loss of the drive system at the current torque distribution point, which consists of copper loss, iron loss, inverter loss, friction loss, stray loss and transmission loss:

$$C_p(T_{mi}) = a_3 T_{mi}^3 + a_2 T_{mi}^2 + a_1 T_{mi} + a_0$$ (15)

where a_3, a_2, a_1 and a_0 are the fitting coefficients, they are fitted out with the experimental data shown in Figure 4.

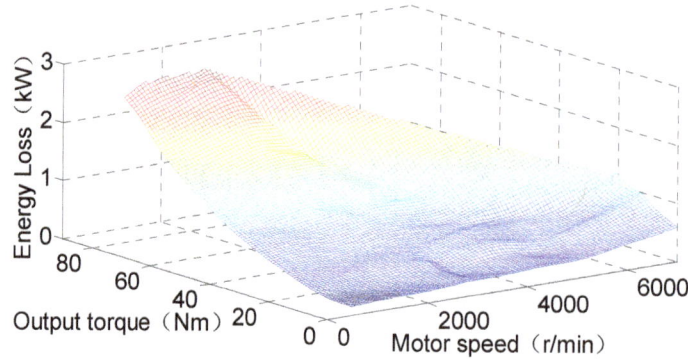

Figure 4. Energy loss *vs.* motor speed and output torque.

In this paper the expression Equation (15) is used to perform the energy-saving control in the penalty function and it is different from that used in [11–14] shown as below:

$$C_p(T_{mi}) = 9549 \sum_{i=1}^{4} \frac{T_{mi} n_i}{\eta(T_{mi})} \qquad (16)$$

In the above Equation (16), n_i is the motor speed and $\eta(T_{mi})$ is a function with relatively high order that will lead to high calculation cost when solving the constructed mathematical programming. Therefore, the expression Equation (15) with relatively low order is used as a part of the penalty function.

The third part of the penalty function is used to put constraint on the wheel slip ratio:

$$C_t(T_{mi}) = \sum_{i=1}^{4} (T_{mi} \lambda_i)^2 \qquad (17)$$

where λ_i is the wheel slip ratio. This item is used for avoiding excessive spin of wheels by reducing the driving torque on the wheels that have relatively high slip ratios.

4.2. Off-Line and On-Line Optimization

As the penalty function in Equation (13) is non-convex and nonlinear, the algorithms used in the literature to search for the global optimal point will lead to heavy calculation burdens and are thus not suitable for practical application. In this paper, the combination of off-line and on-line optimization is used to search for the optimal point of Equation (13), where, off-line optimization based on a simplified penalty function is used to find the optimal point considering yaw moment and energy-saving control, the following on-line optimization is to search for the local optimal point near the off-line result with consideration of the slip ratio constraint. The multi-objective programming is simplified as below, and it is suitable for off-line optimization:

$$\min_{T_{mi}} J_1 = C_p(T_{m1}) + C_p(T_{m3}), \quad \min_{T_{mi}} J_2 = C_p(T_{m2}) + C_p(T_{m4})$$
$$s.t. T_{m1} + T_{m3} = (T_d - M_d / L_s)/2;$$
$$T_{m2} + T_{m4} = (T_d + M_d / L_s)/2; \qquad (18)$$
$$T_{mi\,min} \leq T_{mi} \leq T_{mi\,max}, i = 1, 2, 3, 4.$$

During the simplification, the steering angle is approximated as zero and the slip ratio constraint is neglected. At a certain motor speed the energy loss function $C_p(T_{mi})$ is confirmed, then the problem

Equation (18) can easily be solved by off-line optimization. It is important to note here that in [16], the authors had done some research on the distribution of driving torque between the front axle and rear axle of a four-wheel-drive electric vehicle. In the development of the energy saving-oriented wheel torque distribution controllers, the functional relationship between energy efficiency and drive train output torque is needed. The accuracy of the distribution results must rely on that of the function established. In [16], a second-order function was deduced to describe the relationship between energy loss and drive train output torque based on theoretical analysis. Then the authors drew the conclusion that wheel torque equal division had better effect in vehicle energy saving than control allocation algorithms introduced in the literature. In this paper, the energy efficiency of an electric drive system was gotten based on a dynamometer system, and the functions of energy loss versus output torque are fitted out by the experimental data. In Figure 5, the accuracy of second-order function is compared with the third-order one at different motor speeds.

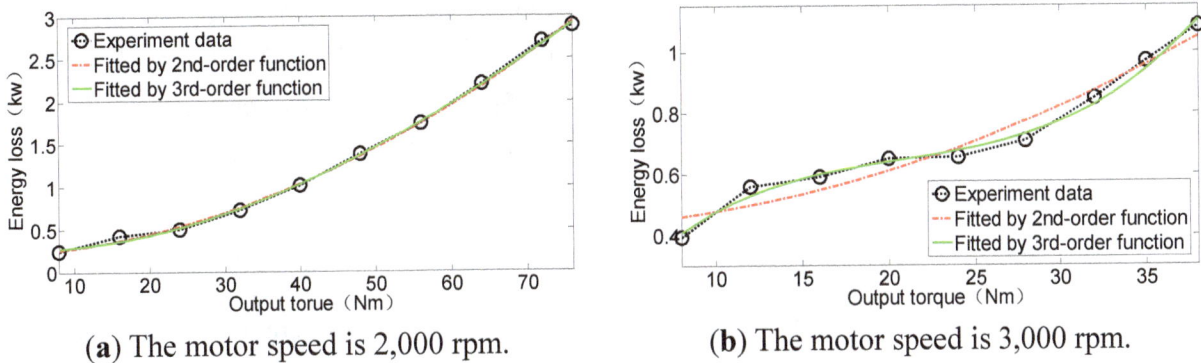

(a) The motor speed is 2,000 rpm.　　　(b) The motor speed is 3,000 rpm.

Figure 5. The fitting function of energy loss *vs.* output torque.

It can be seen from Figure 5 that at 2,000 rpm, both the second-order and third-order functions have high accuracy in the description of the relationship between drive system energy loss and its output torque. At 3,000 rpm, the third-order function is much more accurate than the second-order one. Therefore, in this paper the function adopted in describing the relationship between energy loss and output torque is third-order when the motor speed is 3,000 rpm or above.

Through off-line optimization the partition factor between T_{m1} and T_{m3} or T_{m2} and T_{m4} can be obtained at different motor speed × and torque demand ×, which is shown in Figure 6. We can know that under most conditions including low speed and low desired driving torque conditions, the partition factor between two motors is 0.5. When the motor speed ranges from 3,500 rpm to 7,000 rpm and the desired torque for the two motors from either side of the vehicle is relatively small, the partition factor is 0, which means in this area the desired driving torque should be generated by one of the two motors.

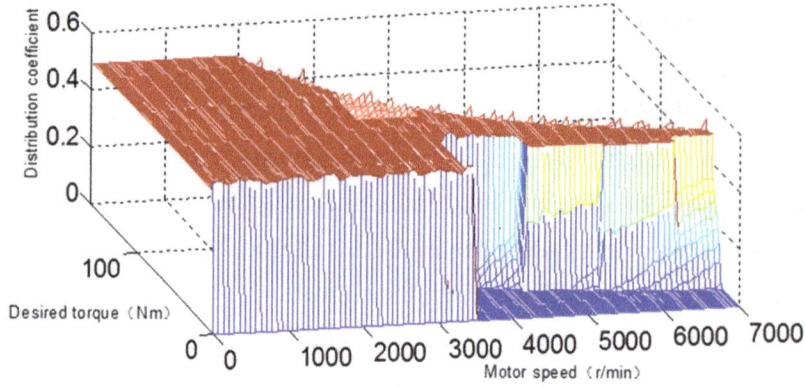

Figure 6. The partition factor based on off-line calculation.

The off-line optimization result above, named T_m^*, will be used as the starting point in the on-line optimization by a 2-D lookup table. One of the inputs of the lookup table is the summation driving torque T_{suml} or T_{sumr} of the two motors in the left or right side of the vehicle, and the other one is the vehicle velocity from which we can get the motor speed. The partial factor shown in Figure 6 is used for the 2-D lookup table. The mentioned summation driving torque T_{suml} and T_{sumr} is figured out by the following expressions:

$$T_{suml} = T_{m1} + T_{m3} = (T_d - M_d / L_s)/2, T_{sumr} = T_{m2} + T_{m4} = (T_d + M_d / L_s)/2$$

In order to take the steering wheel angle and slip ratio constraint into consideration in wheel torque distribution, an on-line optimization step is added to the control allocation process. Newton-Lagrange [17] is used to find the local optimal point around T_m^*. The adopted algorithm can transform the nonlinear and non-convex optimization to quadratic programming at T_m^* shown as below. First, the problem is simplified by removing the numerical limit on wheel torque:

$$\min J(T_m) = (\sum_{i=1}^{4} T_{mi} L_i - V_d)^2 + \sigma_p \sum_{i=1}^{4} C_p(T_{mi}) + \sigma_t C_t(T_{mi}), \; s.t. \; \sum_{i=1}^{4} T_{mi} = T_d . \tag{19}$$

Then the Lagrange function can be written as:

$$L(T_m, \mu) = J(T_m) + \mu h(T_m) \tag{20}$$

where:

$$h(T_m) = \sum_{i=1}^{4} T_{mi} - T_d \tag{21}$$

Thus, the local optimal point around T_m^* can be figured out by the equation as follows:

$$\begin{bmatrix} W(T_m, \mu) & -\nabla h(T_m) \\ -\nabla h(T_m) & 0 \end{bmatrix} \begin{bmatrix} d_k \\ \nu_k \end{bmatrix} = \begin{bmatrix} -\nabla J(T_m) + \nabla h(T_m)\mu_k \\ h(T_m) \end{bmatrix} \tag{22}$$

where, $W(T_m, \mu)$ is the Hessian matrix of the Lagrange function with respect to T_m. d_k and ν_k are the variation of T_m and μ, respectively:

$$W(T_m, \mu) = \nabla_{TT}^2 L(T_m, \mu) = \nabla_{TT}^2 J(T_m) + \mu \nabla_{TT}^2 h(T_m) \tag{23}$$

From Equation (21) we can get:

$$\nabla^2_{TT} h(T_m) = 0 \tag{24}$$

thus:

$$W(T_m, \mu) = \nabla^2_{TT} J(T_m) \tag{25}$$

$$\nabla J(T_m) = \begin{bmatrix} \nabla_1 J(T_m) & \nabla_2 J(T_m) & \nabla_3 J(T_m) & \nabla_4 J(T_m) \end{bmatrix}^T \tag{26}$$

$$\nabla^2_{TT} J(T_m) = \begin{bmatrix} \nabla^2_{11} J(T_m) & \nabla^2_{12} J(T_m) & \nabla^2_{13} J(T_m) & \nabla^2_{14} J(T_m) \\ & \nabla^2_{22} J(T_m) & \nabla^2_{23} J(T_m) & \nabla^2_{24} J(T_m) \\ & & \nabla^2_{33} J(T_m) & \nabla^2_{34} J(T_m) \\ & & & \nabla^2_{44} J(T_m) \end{bmatrix} \tag{27}$$

where:

$$\nabla^2_{ii} J(T_m) = 2L_i^2 + \sigma_p \nabla^2_{ii} C_p(T_m) + 2\sigma_t \lambda_i^2 \tag{28}$$

$$\nabla^2_{ij} J(T_m) = 2L_i L_j \tag{29}$$

therefore:

$$\Delta_T^T W(T_m, \mu) \Delta_T = 2 \left(\sum_{i=1}^4 \Delta_{T_i} L_i \right)^2 + \sum_{i=1}^4 \left(\sigma_p \nabla^2_{ii} C_p(T_m) + 2\sigma_t \lambda_i^2 \right) \Delta_{T_i}^2 \tag{30}$$

We can see from (30), $\forall \Delta_T \neq 0$, $\Delta_T^T W(T_m, \mu) \Delta_T > 0$ can be ensured by properly choosing σ_p and σ_t, so $W(T_m, \mu)$ is a positive definite matrix. Therefore, we can get the solution of Equation (22) by solving the convex optimization problem below:

$$\begin{cases} \min q_k(\Delta) = \dfrac{1}{2} \Delta^T W(T_m^*, \mu^*) \Delta + \nabla J(T_m^*)^T \Delta \\ s.t. \quad h(T_m^*) + \nabla h(T_m^*)^T \Delta = 0. \end{cases} \tag{31}$$

In this paper, the fixed-point algorithm in [18,19] is adopted to solve the problem above. Δ can be used as optimal direction of T_m^*. The optimum step α is figured out by setting a penalty function as follows:

$$P(T_m, \mu) = \| \nabla L(T_m, \mu) \|^2 = \| \nabla J(T_m) + \nabla h(T_m) \mu \|^2 + \| h(T_m) \|^2 \tag{32}$$

$$\alpha = \rho^m \Delta \tag{33}$$

where, $0 < \rho < 1$, m is the smallest non-negative integer that satisfies the following equations:

$$\begin{cases} P(T + \rho^m d, \mu + \rho^m \nu) \leq (1 - \gamma \rho^m) P(T, \mu) \\ T_{\min} \leq T + \rho^m d \leq T_{\max} \end{cases} \tag{34}$$

Then the result of on-line optimization is $T_m^* + \alpha$. The off-line optimization has given out the wheel torque control allocation result considering yaw moment control and energy saving, but without considering the steering angle and slip ratio constraint. In the on-line optimization, the actual steering angle is considered, as well as the slip ratio constraint, which has made the penalty function more complicated. Newton-Lagrange can transform such a nonlinear and non-convex optimization to a convex one at the off-line optimization result T_m^{*}, which will bring down the calculation cost.

5. Simulation and Discussion

Co-simulation based on MATLAB® and Carsim® is carried out to verify the proposed control strategy. The parameters of the target car are shown in Table 1.

Table 1. Vehicle Parameters.

Parameters	Values	Parameters	Values
Vehicle Mass	1486 kg	Motor Number	4
Wheelbase	2.578 m	Motor Rated Power	8 kW
Vehicle Moment of Inertia	2023 kg·m^2	Motor Peak Torque	78 Nm
Wheel Rolling Radius	0.298 m	Motor Rated Speed	2900 rpm

The vehicle performance, provided by equal division of driving torque, off-line optimization based distribution and combination of off-line and on-line optimization control strategy is simulated under two different conditions. Under the simulated driving conditions, the road friction coefficient is 0.85, and the steering wheel angle changes as shown in Figure 7. The acceleration pedal in puts for the two conditions are shown in Figure 8a and Figure 9a.

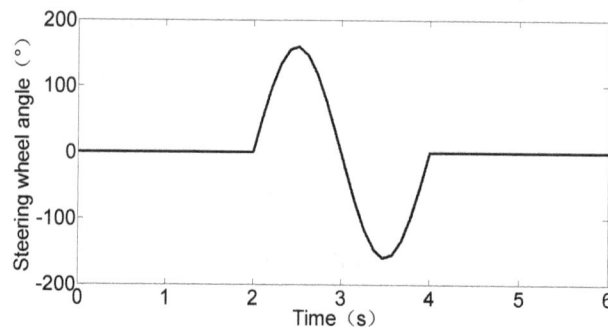

Figure 7. The input of steering wheel during the simulation

5.1. Large Acceleration Simulation

Under this condition, the vehicle initial velocity is set to 20 km/h.

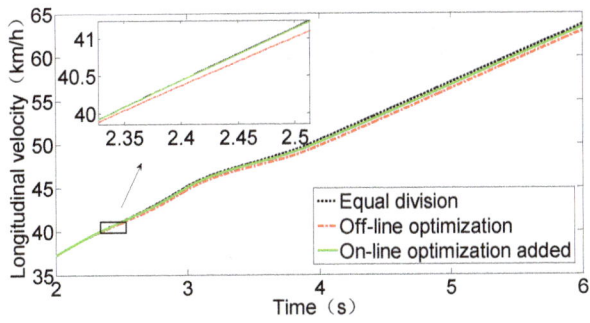

(a) Throttle opening.

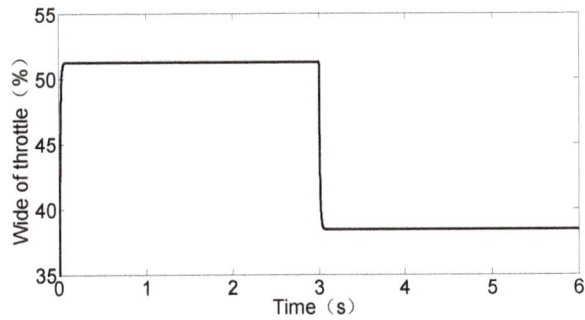

(b) Vehicle longitudinal velocity.

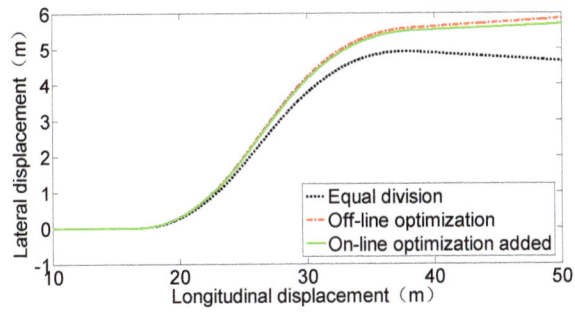

(c) Displacement of the vehicle c.g.

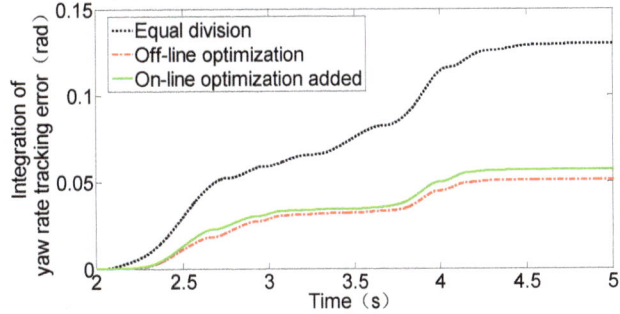

(d) Integration of yaw rate tracking error.

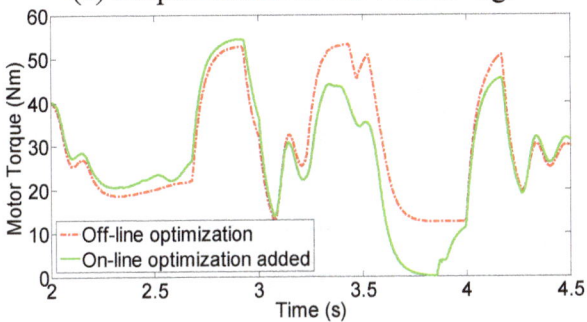

(e) *fl* wheel driving torque.

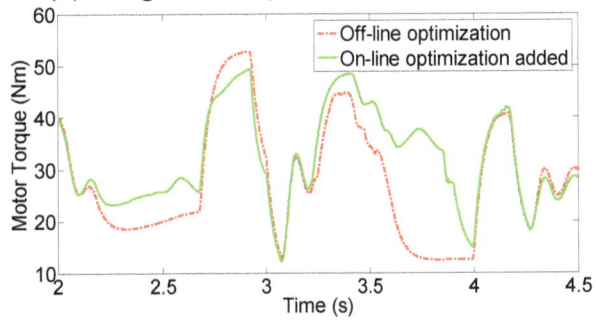

(f) *rl* wheel driving torque.

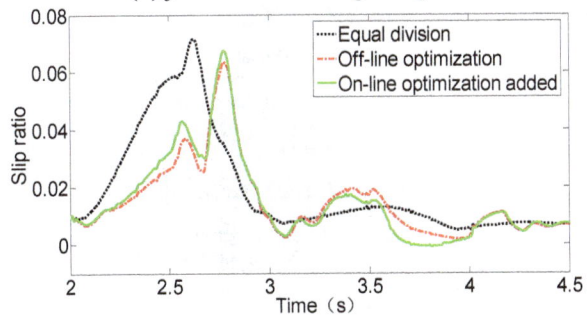

(g) *fl* wheel slip ratio.

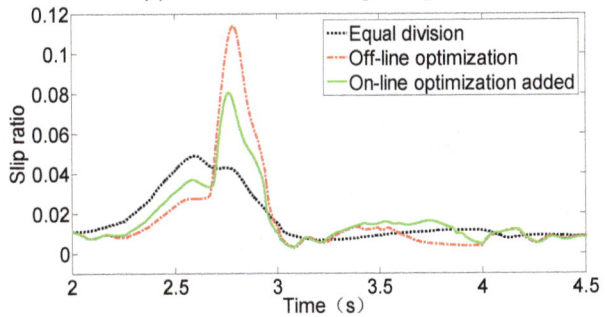

(h) *rl* wheel slip ratio.

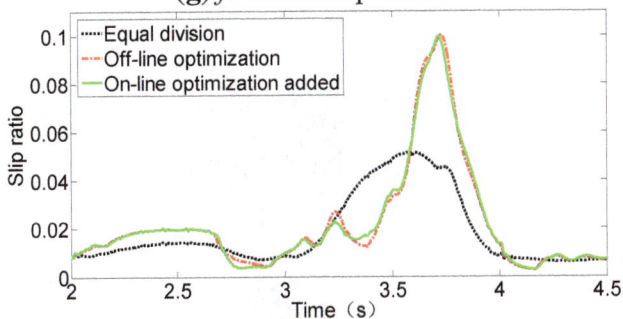

(i) *fr* wheel slip ratio.

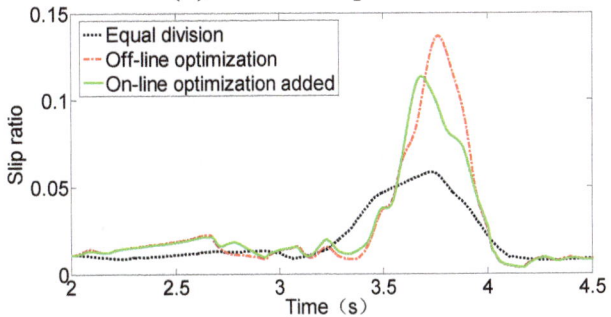

(j) *rr* wheel slip ratio.

Figure 8. *Cont.*

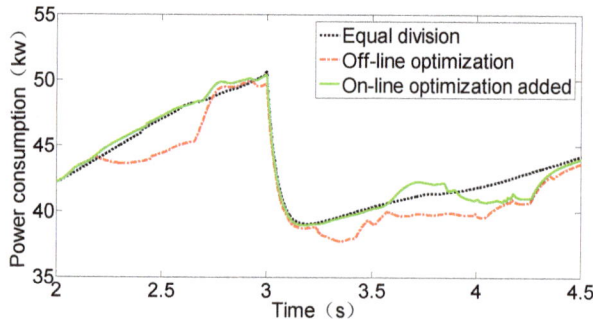

(**k**) Drive system energy consumption.

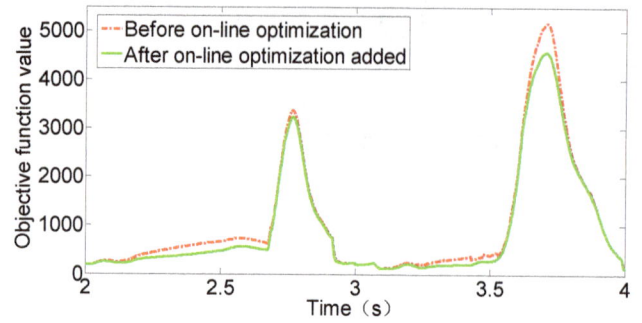

(**l**) Penalty function value.

Figure 8. The vehicle state under large acceleration.

5.2. Small Acceleration Simulation

Under this condition, the vehicle initial velocity is set to 55 km/h.

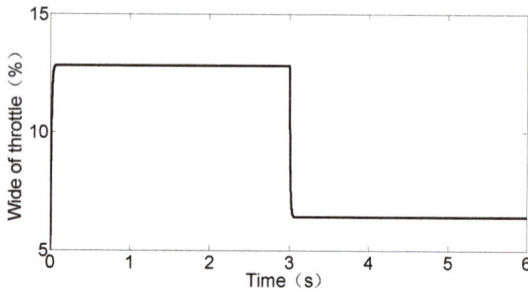

(**a**) Throttle opening.

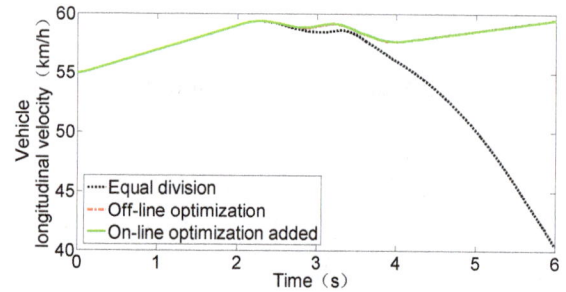

(**b**) Vehicle longitudinal velocity.

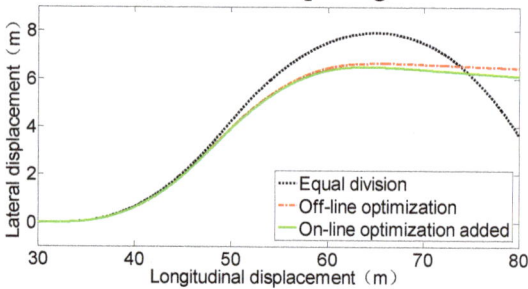

(**c**) Displacement of vehicle c.g.

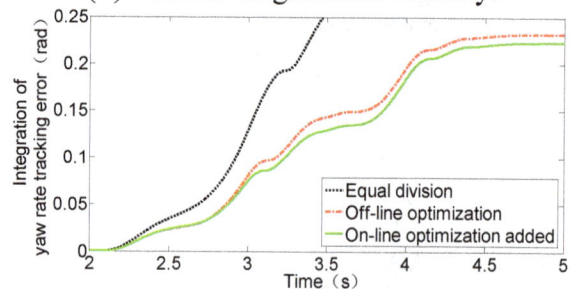

(**d**) Integration of yaw rate tracking error.

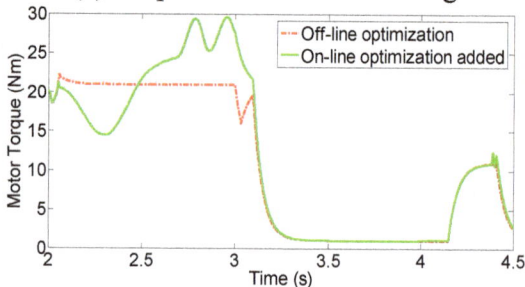

(**e**) *fl* wheel driving torque.

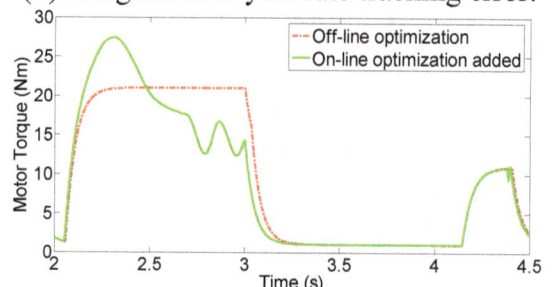

(**f**) *rl* wheel driving torque.

Figure 9. *Cont.*

(g) *fl* wheel slip ratio.

(h) *rl* wheel slip ratio.

(i) *fr* wheel slip ratio.

(j) *rr* wheel slip ratio.

(k) Drive system energy consumption.

(l) Penalty function value.

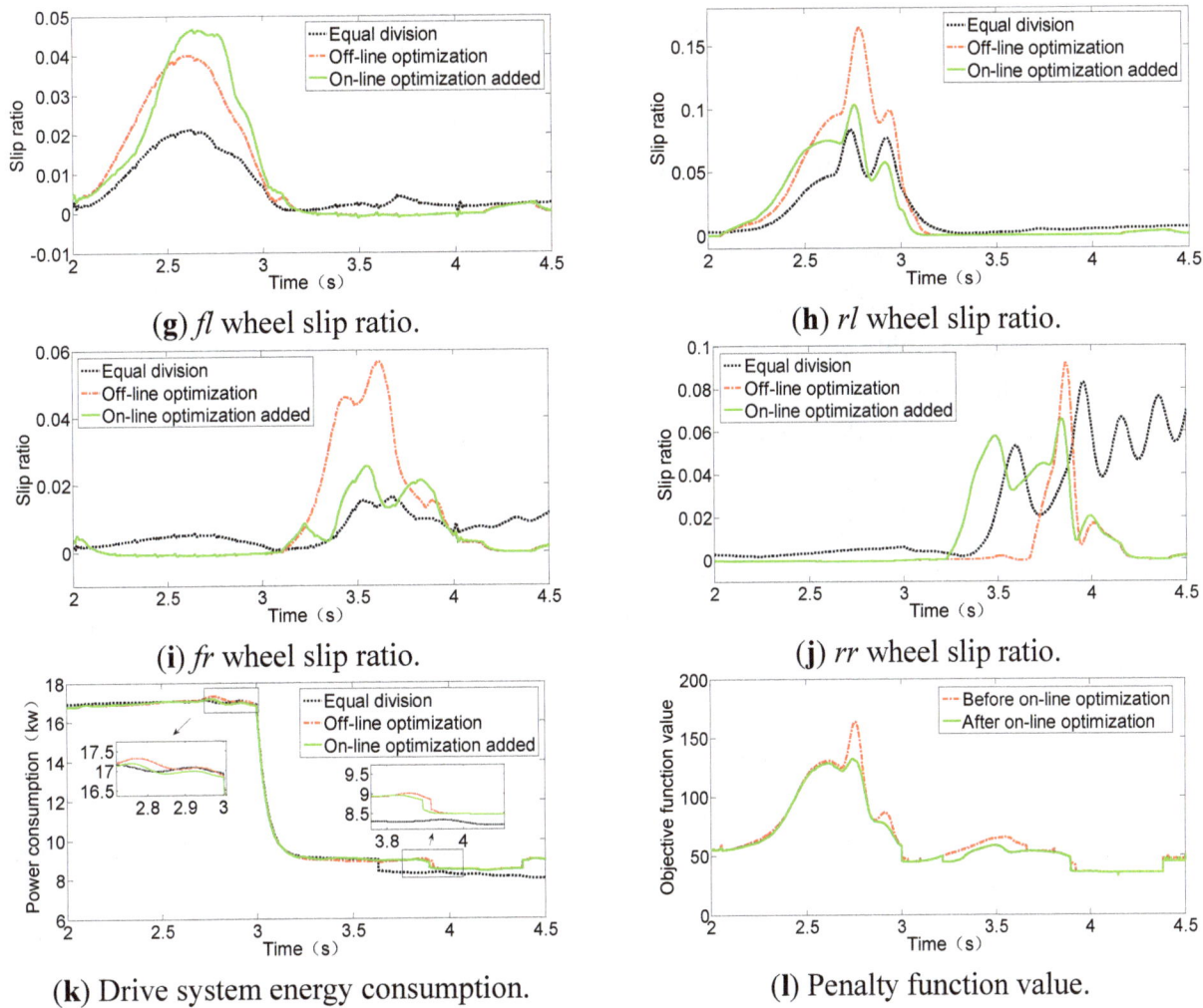

Figure 9. The vehicle state under small acceleration.

5.3. Discussion

Under the combined condition of accelerating and cornering, the wheel torque control strategy should not only meet the demand of vehicle steering maneuverability improvement and energy savings, but also balance the slip ratio of the four wheels to prevent excessive wheel spin. We can know the vehicle is well controlled under the three control strategies from Figure 8b the vehicle velocity time history and 8c the trajectory of the vehicle c.g.. Figure 8d is the integration of yaw rate tracking error, which shows that compared with equal division of wheel torque, either off-line optimization or the combination of off-line and on-line optimization has improved the vehicle tracking ability. This is due to the differential driving performed by the electric drive system when yaw moment is generated. Figure 8e, f are the time history of the *fl* and *rl* wheel driving torque, respectively, from which we can see how wheel torque changes before and after the introduction of on-line optimization. We can see from Figure 8g–j that the slip ratio of rear wheels is reduced after the on-line optimization. Slip ratio constraint added in the penalty function has balanced the slip level among different wheels. Figure 8k is the energy consumption of the drive system under the three different control strategies, from which we can see that when the driving torque is evenly distributed among the four wheels, the energy consumption is the highest, and off-line optimization-based distribution is the most efficient way for saving energy. It is important to note here that, from Figure 6 we know when the motor speed is lower than 3,500 rpm, or car velocity is lower than 53 km/h, wheel even torque distribution is the most efficient method. In Figure 8k, during the first few seconds before the car accelerates to 53 km/h, the power consumption of off-line optimization and evenly distributing control are different, though the wheel driving torque on the left or right side is evenly distributed with off-line optimization. This is because the sum of output torque of off-line optimization is lower than that the driver desires. Figure 8b tells us that the off-line optimization has led to a lower vehicle velocity than evenly distributing control. Off-line optimization generates the biased torque demand between the left and right vehicle side, and it may result in a big torque command that is beyond the motor capability. Therefore, using off-line optimization can reduce energy consumption and the yaw rate tracking error, but deteriorate the vehicle acceleration performance. Figure 8l shows the change of penalty function that has given consideration to the three targets of vehicle maneuverability improvement, energy consumption reduction and slip ratio balance, before and after the on-line optimization.

When the vehicle velocity is high during cornering, a large yaw moment is needed to generate the desired vehicle yaw rate. The longitudinal force of vehicle will lead to reduction of the lateral force that is important for vehicle cornering. Therefore, yaw moment control based on differential driving is especially important when the vehicle is accelerating while cornering with high velocity. From Figure 9b–d we can find that the vehicle has been in a kind of loss control state when wheel driving torque is evenly distributed. Both the off-line optimization and the proposed strategy have successfully controlled the vehicle in the appropriate state. Figure 9e–j also tell us that the on-line optimization has balanced the wheel slip ratios to prevent the excessive spin by reducing wheel driving torque when slip ratio is high. It can be seen from Figure 9k that the energy consumption is diminished when the vehicle is controlled with the proposed strategy, which means the combination of off-line and on-line optimization is better than individual off-line optimization. Figure 9l shows the decrease of the penalty function value after the on-line optimization.

6. Conclusions

The research and application of electric drive technology is an effective way to solve energy and environment issues. Moreover, on 4WDEVs the "fun-to-drive" qualities [20] can be realized by properly distributing wheel torque which may also extend the driving range.

In this paper, a vehicle drive control strategy is developed to allow optimal wheel torque distribution under combined conditions while accelerating and cornering. The proposed approach is developed based on a hierarchical structure. At the high-level the desired driving torque and yaw moment are figured out based on SMC due to the model inaccuracy and parameter error. At the low-level the total driving torque is allocated to wheels by multi-objective optimization. The slip ratio constraint has been added to the penalty function to prevent excessive spin. Another two parts of the penalty function are yaw moment control offset and energy loss instead of energy consumption applied in the literature. In this way the penalty function is simplified which makes the calculation cost reduced. The multi-objective programming is solved based on a combination of off-line optimization and on-line optimization algorithm. At the first process, off-line optimization is carried out based on the simplified penalty function, and the result has been used as the start point of on-line optimization. During the on-line optimization, local minimal point is reached around the off-line result. Co-simulation based on

MATLAB® and Carsim® has verified the presented wheel torque distribution process in terms of both maneuverability control and energy savings. The verification based on vehicle test will be carried out after a 4WDEV is developed.

Acknowledgments

This work was supported by National Natural Science Foundation of China (no.51175043).

Author Contributions

Lin Cheng and Zhifeng Xu built the dynamic model of vehicle dynamics control for 4WDEV, developed the wheel torque control strategy and performed the simulations.

Conflicts of Interest

The authors declare no conflict of interest.

References

1. Wu, T.; Zhang, M.B.; Ou, X.M. Analysis of future vehicle energy demand in China based on a Gompertz function method and computable general equilibrium model. *Energies* **2014**, *7*, 7454–7482.
2. He, H.W.; Peng, J.K.; Xiong, R.; Fan, H. An acceleration slip regulation strategy for four-wheel drive electric vehicles based on sliding mode control. *Energies* **2014**, *7*, 3748–3763.
3. Sakai, S.; Sado, H.; Hori, Y. Motion control in an electric vehicle with four independently driven in-wheel motors. *IEEE/ASME Trans. Mechatron.* **1999**, *4*, 9–16.
4. Lin, C.; Zhang, Z.J.; Ma, J. Dual-motor Anti-slip Differential Drive System. Chinese Patent China 200810097693.5, 2009.
5. Jeongmin, K.; Chiman, P. Control algorithm for an independent motor-drive vehicle. *IEEE Trans. Veh. Technol.* **2010**, *59*, 3213–3222.
6. Li, F.Q.; Wang, J.; Liu, Z.D. Motor torque based vehicle stability control for four-wheel-drive electric vehicle. In Proceedings of the IEEE Vehicle Power and Propulsion Conference, Dearborn, MI, USA, 7–11 September 2009; pp. 1596–1601.
7. Peng, H.; Hori, Y. Optimum traction force distribution for stability improvement of 4WDEV in critical driving condition. In Proceedings of the 9th IEEE International Workshop on Advanced Motion Control, Istanbul, Turkey, 27–29 March 2006; pp. 596–601.
8. He, Z.Y.; Ji, X.W. Nonlinear robust control of integrated vehicle dynamics. *Veh. Syst. Dyn.* **2012**, *50*, 247–280.
9. Falcone, P.; Tseng, H.E.; Borrelli, F.; Asgari, J.; Hrovat, D. MPC-based yaw and lateral stabilization via active front steering and braking. *Veh. Syst. Dyn.* **2008**, *46*, 611–628.
10. Javad, A.; Ali, K.S.; Mansour, K. Adaptive vehicle lateral-plane motion control using optimal tire friction forces with saturation limits consideration. *IEEE Trans. Veh. Technol.* **2009**, *58*, 4098–4107.

11. Wang, R.R.; Chen, Y.; Feng, D.W. Development and performance characterization of an electric ground vehicle with independently actuated in-wheel motors. *J. Power Sources* **2011**, *196*, 3962–3971.

12. Chen, Y.; Wang, J.M. Energy-efficient control allocation with applications on planar motion control of electric ground vehicles. In Proceeding of the 2011 American Control Conference, San Francisco, CA, USA, 29 June–1 July 2011; pp. 2719–2724.

13. Chen, Y.; Wang, J.M. Design and experimental evaluations on energy efficient control allocation methods for over actuated electric vehicles: Longitudinal motion case. *IEEE/ASME Trans. Mechatron.* **2014**, *19*, 538–548.

14. Leonardo, D.N.; Aldo, S.; Patrick, G. Wheel torque distribution criteria for electric vehicles with torque-vectoring differentials. *IEEE Trans. Veh. Technol.* **2014**, *63*, 1593–1602.

15. Horiuchi, S.; Okada, K.; Nohtomi, S. Improvement of vehicle handling by nonlinear integrated control of four wheel steering and four wheel torque. *JSAE Rev.* **1999**, *20*, 459–464.

16. Gu, J.; Ouyang, M.; Lu, D. Energy efficiency optimization of electric vehicle driven by in-wheel motor. *Int. J. Automot. Technol.* **2013**, *14*, 763–772.

17. Huang, H.X.; Han, J.Y. *Mathematical Programming*; Tsinghua University Press: Beijing, China, 2006; pp. 289–291.

18. Lu, P. Constrained tracking control of nonlinear systems. *Syst. Control Lett.* **1997**, *27*, 305–314.

19. Wang, J.M. *Coordinated and Reconfigurable Vehicle Dynamics Control*; The University of Texas at Austin: Austin, TX, USA, 2007; pp. 80–104.

20. Novellis, L.D.; Sorniotti, A.; Gruber, P. Optimal wheel torque distribution for a four-wheel-drive fully electric vehicle. *SAE Int. J. Passeng. Cars–Mech. Syst.* **2013**, *6*, 128–136.

8

Comparative Study of a Fault-Tolerant Multiphase Wound-Field Doubly Salient Machine for Electrical Actuators

Li-Wei Shi * and Bo Zhou

Aero-Power Science Technology Center, Nanjing University of Aeronautics and Astronautics, Nanjing 210016, China; E-Mail: zb3713@gmail.com

* Author to whom correspondence should be addressed; E-Mail: liwei10@nuaa.edu.cn

Academic Editor: Paul Stewart

Abstract: New multiphase Wound-Field Doubly Salient Machines (WFDSMs) for electrical actuators with symmetric phases are investigated and compared in this paper. With a comparative study of the pole number and pole arc coefficient, the salient pole topology of the three-phase, four-phase, five-phase, and six-phase WFDSMs with little cogging torque is presented. A new winding configuration that can provide symmetrical phases for the multiphase WFDSMs is proposed. Suitable fault-tolerant converters for the multiphase WFDSM are presented. With the simulated results in terms of the pole topology, flux linkage, back EMF and converters, it can be concluded that the pole numbers of the new five-phase WFDSM are very large. The high accuracy position sensors should be required to make the five-phase WFDSM commutate frequently and accurately at a high speed. The four-phase and the six-phase WFDSM can be divided into two isolated channels, and both of them have a good performance as a fault-tolerant machine. All of the investigations are verified by finite element analysis results.

Keywords: wound-field doubly salient machine; electrical actuation; fault-tolerant; multiphase

1. Introduction

Today conventional aircraft are characterized by complex hydraulic nets. In order to reduce the weight of the pipelines, cylinders, pump, valves and switches of the hydraulic system, the aircraft is adopting more and more electrical systems in preference to others. Now, researchers and engineers have proved that electrical actuators can be used to reduce or to remove the traditional hydraulic, and mechanical systems in the next few years [1]. The more electric aircraft approach is widely discussed in the technical literature, which includes the following three main drives [2]:

- The starter-generator for the engine;
- The electrical actuators for the flight control;
- The electric machines for the fuel pump.

There are many different types of actuators in a conventional aircraft [3], such as the actuators in the wings and in the tail. In the hydraulic actuation system, the flight control is realized by a hydraulic pump and a hydraulic motor, several fluid pipelines and hydraulic actuators. Now, more and more electric machines are being used to replace or assist the hydraulic actuation system. For example, in the Boeing 787, the spoilers and the horizontal stabilizer flight controls are driven by electric machines in order to guarantee the operation in the case of a hydraulic failure.

A literature review reveals that several types of machine can be used as a drive motor for electrical actuators [4]. Among them, PM machines and Switched Reluctances Machines (SRMs) were abundantly studied in the past years, because of their very-high power density.

In [5], a five-phase PM brushless machine was developed for an aircraft flap actuator application, and the machine can endure the fault of one or two open phases or a phase short circuit. In [6], a PM fractional slot machine was designed, because the fractional slot windings have low mutual inductances between phases, which meet the magnetic isolation demands of phase windings for multiphase fault-tolerant PM synchronous machines [7]. Such fault-tolerant PM machines were also studied in [8,9].

It is necessary to remark that the actuators of an airplane have to work in very harsh ambient conditions, with temperature variations from −60 °C to +70 °C and the air pressure varies from almost 0 to 1 bar [2]. This harsh environment puts forward higher requirements for high performance PM materials. Furthermore, many of the electrical actuators care little about the torque ripple because the noise of the airplane is very high. Therefore, switched reluctances machines were investigated to drive the actuators in [10,11].

The Wound-Field Doubly Salient Machine (WFDSM) has the same rotor as the SRM that will not suffer from faults of the PM materials or brush faults of the wound-field synchronous motors. The WFDSM is derived from a doubly salient PM machine (DSPM) [12] by using field windings instead of permanent magnet excitation [13]. The WFDSM provides the excitation flux by the DC field windings instead of the PMs. Therefore, the output torque and speed can be adjusted by the field winding and phase windings. What's more, the phase windings of the WFDSM are isolated from each other, and it has low mutual inductances between phases, which reduces the negative influence of the faulty phase. It has broad application prospects in the fields that care little about the torque ripple, such as mining machinery, electrical actuators, and starter-generators.

For example, a WFDSM with two-section twisted-rotor was developed as a starter-generator for aerospace applications [14]. Recently, some new three-phase WFDSMs with new winding arrangements have been developed to take the place of traditional electric machines [15,16]. In the energy conversion area, a WFDSM worked as a DC generator was equipped in an EV range extender [17]. A prototype of a 24/32-pole WFDSM was developed as a low speed wind turbine generator [13].

Multiphase machines with more than three phases can be applied for high reliability applications because they can still run even with one or two open-circuited phases [18]. To improve the reliability of the WFDSM, a traditional four-phase WFDSM was studied in [17], which was similar to the 8/6-pole DSPM described in [19]. A five-phase WFDSM was developed as a generator in [20], which showed that it had good fault-tolerant characteristics. Therefore, the multiphase WFDSM is very suitable to be designed as a fault-tolerant machine.

However, the four-phase and the five-phase WFDSMs discussed above are traditional WFDSMs with their field windings wound around four and five stator poles, respectively. They have the disadvantage of phase asymmetry. The phase asymmetry of the three-phase WFDSM is not obvious, but we found that the phase asymmetry will increase with the number of phases. It was considered that the WFDSM cannot be designed with six phases, and there have been no reports of six-phase WFDSMs until now.

In this paper, new multiphase WFDSMs for electrical actuators with symmetric phases will be investigated and compared. With the comparative study of the poles number and pole arc coefficient, the salient pole topology of the WFDSMs that have little cogging torque will be presented. A new winding configuration to provide symmetrical phases will be proposed. Suitable fault-tolerant converters for the multiphase WFDSMs will be presented. With the comparison in terms of the pole topology, flux linkage, back EMF and converters, comparative conclusions will be proposed to select a multiphase WFDSM for electrical actuators.

2. Comparative Study of the Salient Pole Topology

2.1. Salient Pole Number

There is a wide range of possible combinations of the stator poles and the rotor poles. Nevertheless, only few combinations are suitable to be selected. To outline the pole combinations, the principle of the salient pole number should be studied first.

As we can see from the traditional three-phase WFDSM in Figure 1a, there are $6N$ stator poles and $4N$ rotor poles, where N is the number of element machines. Each phase coil is wound around one stator pole, and the field coils are wound around every three stator poles [17].

So the first law of the stator poles can be written as:

$$p_s = mi \tag{1}$$

where p_s should be an even number, and it stands for the number of stator poles. m is the phase number and i is a positive integer. If p_s is an odd number, the north field winding will not be equal to the south field winding, and the machine will generate an unbalanced magnetic force. Let p_r be the rotor poles number. The mechanical angle of one period β_r can be obtained as:

$$\beta_r = \frac{360°}{p_r} \tag{2}$$

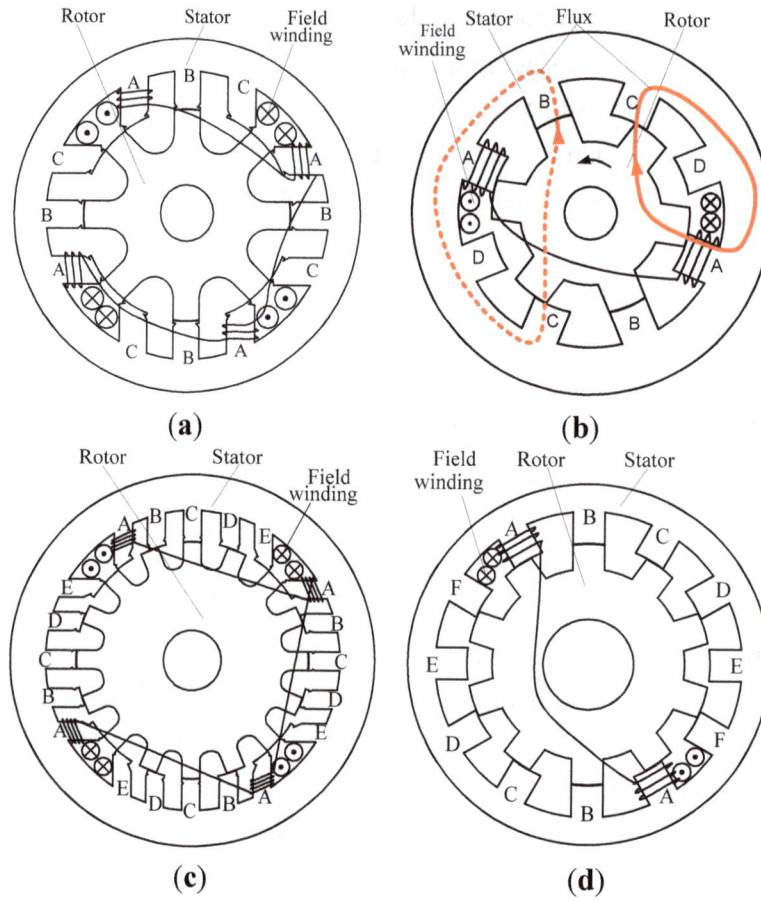

Figure 1. Structure of the three-phase and multiphase DSG. (**a**) 12/8-pole three-phase of WFDSM; (**b**) 8/6-pole four-phase; (**c**) 20/16-pole five-phase; (**d**) 12/10-pole six-phase.

The mechanical angle of each stator pole β_s is:

$$\beta_s = \frac{360°}{p_s} \tag{3}$$

Because the adjacent poles have the adjacent phase windings, the mechanical angle between two phases can be expressed as:

$$\beta_\delta = \frac{360°}{m} \cdot \frac{1}{p_r} \tag{4}$$

and the difference between β_r and β_s is $\pm\beta_\delta$:

$$\beta_r - \beta_s = \pm\beta_\delta \tag{5}$$

From Equations (2)–(5), we can get the second law of the rotor and the stator poles:

$$\frac{p_s}{p_r} = \frac{m}{m \pm 1} \tag{6}$$

When $m = 3$, the elementary machine of three-phase WFDSM has six stator poles and four or eight rotor poles, which is called 6/4-pole machine or 6/8-pole machine [14]. In the same way, the elementary machine of a traditional four-phase WFDSM has an 8/6-pole or 8/10-pole structure. Table 1 gives the pole combinations of the WFDSM with different phases.

Table 1. The poles combinations of the WFDSM.

Phase number	Stator poles	Rotor poles	Example
Three-phase	6N	4N or 8N	12/8
Four-phase	4N	3N or 5N	8/6
Five-phase	10N	8N or 12N	20/16
Six-phase	6N	5N or 7N	12/10

2.2. Pole Arc Coefficient

The WFDSM stator is equipped with both field coils and phase coils. The self-inductance of the phase winding and the field winding change with the rotor positions and the pole arcs [21,22]. If the pole arc is not well-designed, the machine will generate torque ripples because the reluctance and the flux of the field winding will change with the rotor position [23]. In order to minimize the cogging torque caused by the mutative reluctance of the field winding, the self-inductance of the field winding should be constant when the rotor rotates. Overall, the increasing phase number should be equal to the decreasing phase number.

For the common inner-rotor motor, the rotor pole number is usually less than the stator pole number. Hence the rotor pole is generally wider than the stator pole. As the narrow pole of the stator and the rotor determines the increasing or decreasing mechanical angle of the phase inductance, the increasing mechanical angle of the phase inductance can be described as:

$$\beta_{working} = \frac{360°}{p_s} \cdot \alpha_s \tag{7}$$

where α_s is the stator pole arc coefficient, which is the proportion of the stator pole arc length l_t and the pole pitch l_p.

$$\alpha_s = \alpha_s = \frac{l_t}{l_p} \tag{8}$$

Let the electrical angle of one phase voltage waveform be θ:

$$\theta = \frac{360°}{p_s} \alpha_s p_r = x \frac{180°}{m} \tag{9}$$

where x is the phase number that has a mutative self-inductance at any time, and $x \le m$. If x is large, there will be more phases that can output torque or voltage, and the fault-tolerant ability of the machine will be strong. Because the phase number with increasing inductance should be equal to the phase number with decreasing inductance, it can be concluded that x should be an even number.

For the three-phase WFDSM, while $p_s/p_r = 3/2$, $\theta = 120°$. We can draw from Equation (7) and Equation (8) that:

$$\alpha_s = \frac{p_s}{3 p_r} \tag{10}$$

Therefore, to make the machine have no cogging torque, reluctance and flux of the field winding the pole arc coefficient of the three-phase WFDSM α_s is equal to 0.5.

While $p_s > p_r$, the rotor pole width is not generally thinner than the stator pole width. For the three-phase WFDSM, the mutual inductances of the field winding and the phase windings L_{pf} are shown in Figure 2. When $\alpha_r = 0.5$, the machine can be easily controlled by a BLDC controller, because one period can be divided into six equal parts like a BLDC machine. When $\alpha_r = 0.333$, the machine can output a large torque because the rotor poles and the stator poles are monospaced and the leakage flux is small. Overall, the stator and the rotor pole arc coefficient should comply with $\alpha_s = 0.5$ and $\alpha_r = 0.5$ or 0.333.

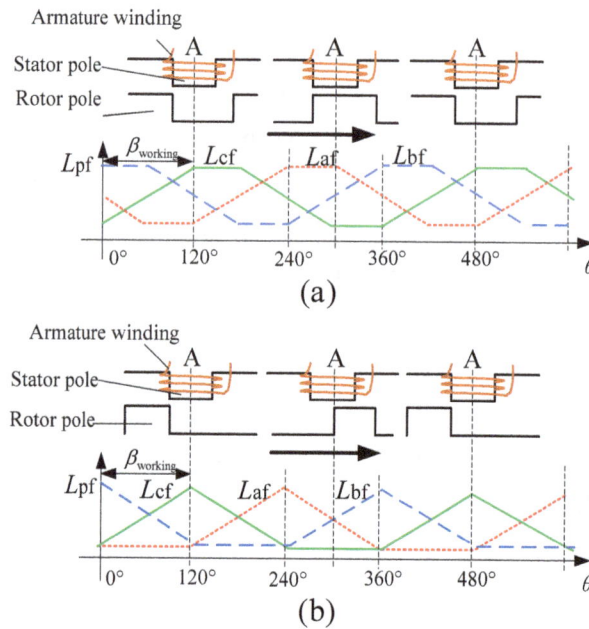

Figure 2. L_{pf} of the three-phase WFDSM with different α_r. (**a**) $\alpha_r = 0.333$; (**b**) $\alpha_r = 0.5$.

It can be concluded that the pole arc coefficients of three-phase WFDSM should comply with Equation (11):

$$\left.\begin{array}{l} \dfrac{p_s}{p_r} = \dfrac{3}{2}: \ \alpha_s = 0.5; \ \alpha_r = 0.333 \text{ or } 0.5 \\[4mm] \dfrac{p_s}{p_r} = \dfrac{3}{4}: \ \alpha_s = 0.25; \ \alpha_r = 0.333 \text{ or } 0.5 \end{array}\right\} \tag{11}$$

However, the four-phase WFDSM should not be designed with $\alpha_r = 0.5$. As we can see from Figure 3, if $\alpha_r = 0.5$, there will be three changing inductances at any time, $x = 3$. With this structure, the reluctance of the field winding will change with the position of the rotor, which will generate a big cogging torque, as well as field winding back EMF.

In order to solve the problem of changeable field reluctance, a new four-phase WFDSM is proposed, whose pole arc coefficient complies with Equation (15):

$$\left.\begin{array}{l} \dfrac{p_r}{p_s}=\dfrac{3}{4}:\ \alpha_s=0.667;\ \alpha_r=0.5 \\[4mm] \dfrac{p_r}{p_s}=\dfrac{5}{4}:\ \alpha_s=0.4;\ \alpha_r=0.5 \end{array}\right\} \tag{12}$$

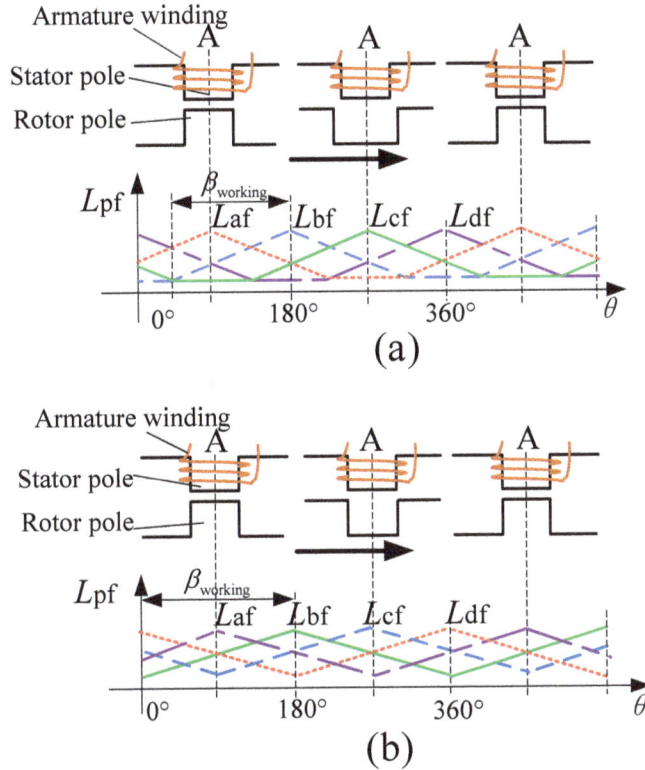

Figure 3. L_{pf} of the four-phase WFDSM with different α_r. (**a**) $\alpha_s = 0.5$, $\alpha_r = 0.375$; (**b**) $\alpha_s = 0.667$, $\alpha_r = 0.5$.

For this new machine, there are two increasing phase inductances and two decreasing phase inductances at any time, as shown in Figure 3b. The field inductance is steady, and it will not generate cogging torque. The phase voltage waveform electrical angle of this machine is 180°, and the phase number with mutative inductance at any time $x = 4$. Therefore, all of the four-phase windings can output torque at any time, which improves the fault tolerance of the machine.

Similarly, we can also deduce the pole arc coefficient of the other multiphase WFDSM, since the phase voltage waveform electrical angle of the five-phase WFDSM is 144°, and the angle of the six-phase WFDSM is 120°. In short, it can be newly concluded that the pole arc coefficient of the multi-phase WFDSM should comply with Equation (13).

In short, to reduce the torque ripple caused by the field winding of multi-phase WFDSM, the stator poles, rotor poles and pole arc should follow topology criteria as shown in Equations (1), (6) and (13).

The three pieces of topology criteria not only give a design basis for the WFDSMs with less than six phases, they can also be used to design PM doubly salient machines and hybrid excitation doubly salient machines. What is more, these topology criteria also provide a derivation example for the WFDSMs with more than seven phases.

$$\left.\begin{array}{llll}\text{three}-\text{phase}, & \dfrac{p_s}{p_r}=\dfrac{3}{2}, & \alpha_s=0.5; & \alpha_r=0.333 \text{ or } 0.5 \\[3mm] \text{three}-\text{phase}, & \dfrac{p_s}{p_r}=\dfrac{3}{4}, & \alpha_s=0.25; & \alpha_r=0.333 \text{ or } 0.5 \\[3mm] \text{four}-\text{phase}, & \dfrac{p_s}{p_r}=\dfrac{4}{3}, & \alpha_s=0.667; & \alpha_r=0.5 \\[3mm] \text{four}-\text{phase}, & \dfrac{p_s}{p_r}=\dfrac{4}{5}, & \alpha_s=0.4; & \alpha_r=0.5 \\[3mm] \text{five}-\text{phase}, & \dfrac{p_s}{p_r}=\dfrac{5}{4}, & \alpha_s=0.5; & \alpha_r=0.4 \\[3mm] \text{five}-\text{phase}, & \dfrac{p_s}{p_r}=\dfrac{5}{6}, & \alpha_s=0.333; & \alpha_r=0.4 \\[3mm] \text{six}-\text{phase}, & \dfrac{p_s}{p_r}=\dfrac{6}{5}, & \alpha_s=0.4; & \alpha_r=0.333 \\[3mm] \text{six}-\text{phase}, & \dfrac{p_s}{p_r}=\dfrac{6}{7}, & \alpha_s=0.571; & \alpha_r=0.333 \end{array}\right\} \qquad (13)$$

Because the pole numbers and the switches of the converter increase with the phase number, the WFDSM with more than six phases is not suitable to be applied because of the weight and the cost of the converter is unacceptable, as well as the pole number. Therefore, their application prospects are not as broad as those of WFDSMs with less than six phases, because they are too complicated. This paper focuses on the WFDSMs with less than seven phases.

2.3. Simulation Results

Figure 4 shows L_{pf} and back EMF waveforms of the traditional multiphase WFDSMs, which are 12/8-pole three-phase, 8/6-pole four-phase 10/8-pole five-phase and 12/10-pole six-phase with the same stator pole arc coefficient $\alpha_s = 0.5$, and the rotor poles are as wide as the stator poles. The simulated result in Figure 4a shows that L_{pf} of the three-phase and four-phase WFDSM are consistent with the analysis result in Figures 2 and 3. The back EMF waveforms in Figure 4b show that the electrical angles of the phase voltage waveforms of the four machines are approximately 120°, 135°, 144° and 150°. This verifies the calculation results in Equation (8).

With the total torque formula given in Equation (14), we can see that there is a torque component $\dfrac{1}{2}i_f^2\dfrac{dL_f}{d\theta}$ which has nothing to do with the phase current i_p. It can be called a cogging torque.

$$T=\frac{1}{2}i_p^2\frac{dL_p}{d\theta}+i_p i_f\frac{dL_{pf}}{d\theta}+\frac{1}{2}i_f^2\frac{dL_f}{d\theta} \qquad (14)$$

where i_p and i_f are the phase current and field current. In order to reduce the negative impact of the armature reaction, the number of turns of the field winding is much larger than the number of phase windings.

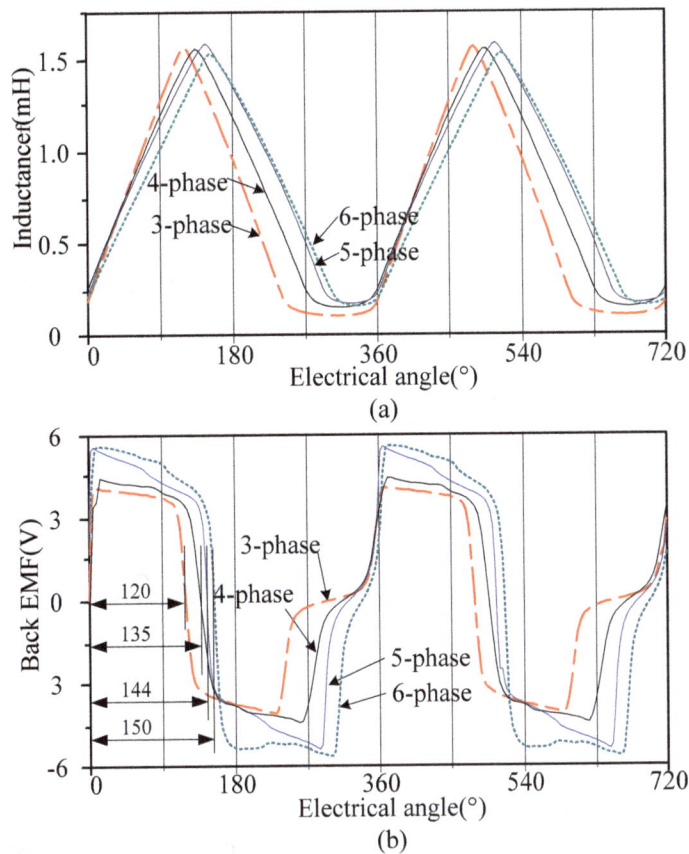

Figure 4. L_{pf} and u_p of multi-phase WFDSM. (**a**) L_{pf}; (**b**) The back EMF.

The self-inductance of the field winding L_f is much larger than the mutual inductance between the phase winding and field winding L_{pf}. Therefore, if L_f changes with the rotor position, and machine will generate a large cogging torque.

For the four-phase WFDSM with $\alpha_s = 0.5$, $\theta = 135°$, and $x = 3$, there are three inductances changing at any time. This situation does not conform to Equation (13). With the self-inductance of the field winding L_f waveform in Figure 5a, we can see that L_f will change with the rotor position, which will produce cogging torque ripples. If $\alpha_s = 0.667$ and $x = 4$, the self-inductance of the field winding L_f will be a constant, as shown in Figure 5b. Therefore, if x is an odd number, the machine will not generate cogging torque.

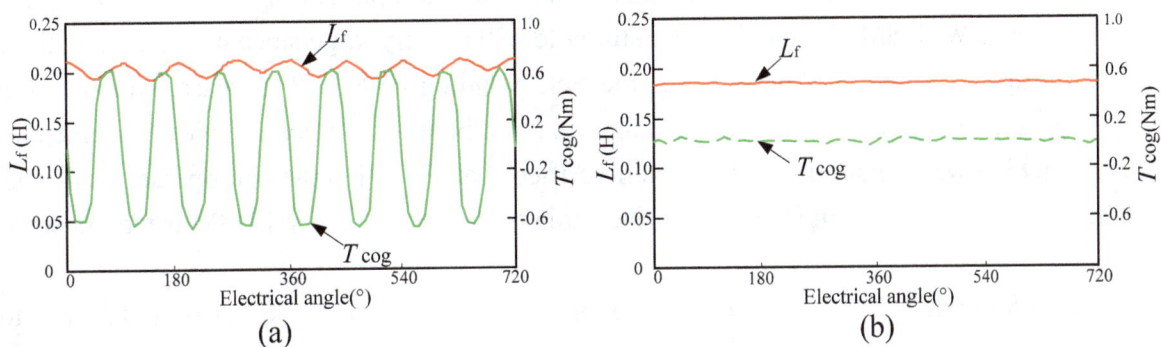

Figure 5. L_f of the four-phase WFDSM. (**a**) $\alpha_s = 0.5$; (**b**) $\alpha_s = 0.667$.

3. Comparative Study of the Symmetrical Phase Winding Configuration

3.1. The Symmetrical Phase Winding Configuration

As shown in Figure 1a, there are four field coils that are used to provide the magnetic field. Each field coil is wound around three stator poles. In Figure 1b, there are four stator poles in a field coil, which provides the magnetic field. Therefore, each excitation source of traditional m-phase doubly salient machine couples with m-phase coils. As the red lines show in Figure 1b, the magnetic circuit of phase A and D which are close to the excitation source is much shorter than phase B and C, and the inductance of phase A and D are larger than that of phase B and C. Overall the amplitudes of the inductance of the traditional four-phase WFDSM have the relationship given by Equation (15):

$$\max(L_{af}) = \max(L_{df}) > \max(L_{bf}) = \max(L_{cf}) \tag{15}$$

If the phase coils with short flux road are divided averagely, the total inductances added by the series coils will be equal [12]. Let j stand for the number of phase coils that coupled by a field coil. The stator poles can be calculated with:

$$P_s = m \cdot k = jP \tag{16}$$

where k and P are two natural numbers. Therefore, P_s is the least common multiple of m and j at least. Together with Equation (6), we can list the pole numbers of the four-phase WFDSM with different j in Table 2.

Table 2. The pole numbers of the four-phase WFDSM with different j.

j	Poles of an element machine		
	Four-phase	Five-phase	Six-phase
$j = 1$	8/6	10/8	12/10
$j = 2$	8/6	10/8	12/10
$j = 3$	12/9	30/24	12/10
$j = 4$	8/6	40/32	24/20
$j = 5$	–	10/8	30/25
$j = 6$	–	–	12/10

When $j = m$, the machine is a traditional WFDSM with asymmetric phases. If $j = 4$ in the five-phase WFDSM and six-phase WFDSM, the number of stator pole will be very large since it increases with the least common multiple of j and m. And if $j = 2$, the field winding coils will increase, which in turn increases the copper consumption of the field winding. A three-phase 6/4-pole variable flux reluctance doubly salient machine was reported in [16], which verified that the WFDSM can operate well when $j = 1$. With the same configuration in [16], every stator pole of the multiphase WFDSM is wound with a field winding.

In the traditional 8/6-pole WFDSM, as shown in Figure 6a, each field coil is wound around four stator poles. All the coils of phase A and phase D are nearby the field coil slots, and the coils of the other two phases are in the middle of the two field coil slots. Therefore, the total reluctance of phase B is larger than the reluctance of phase A.

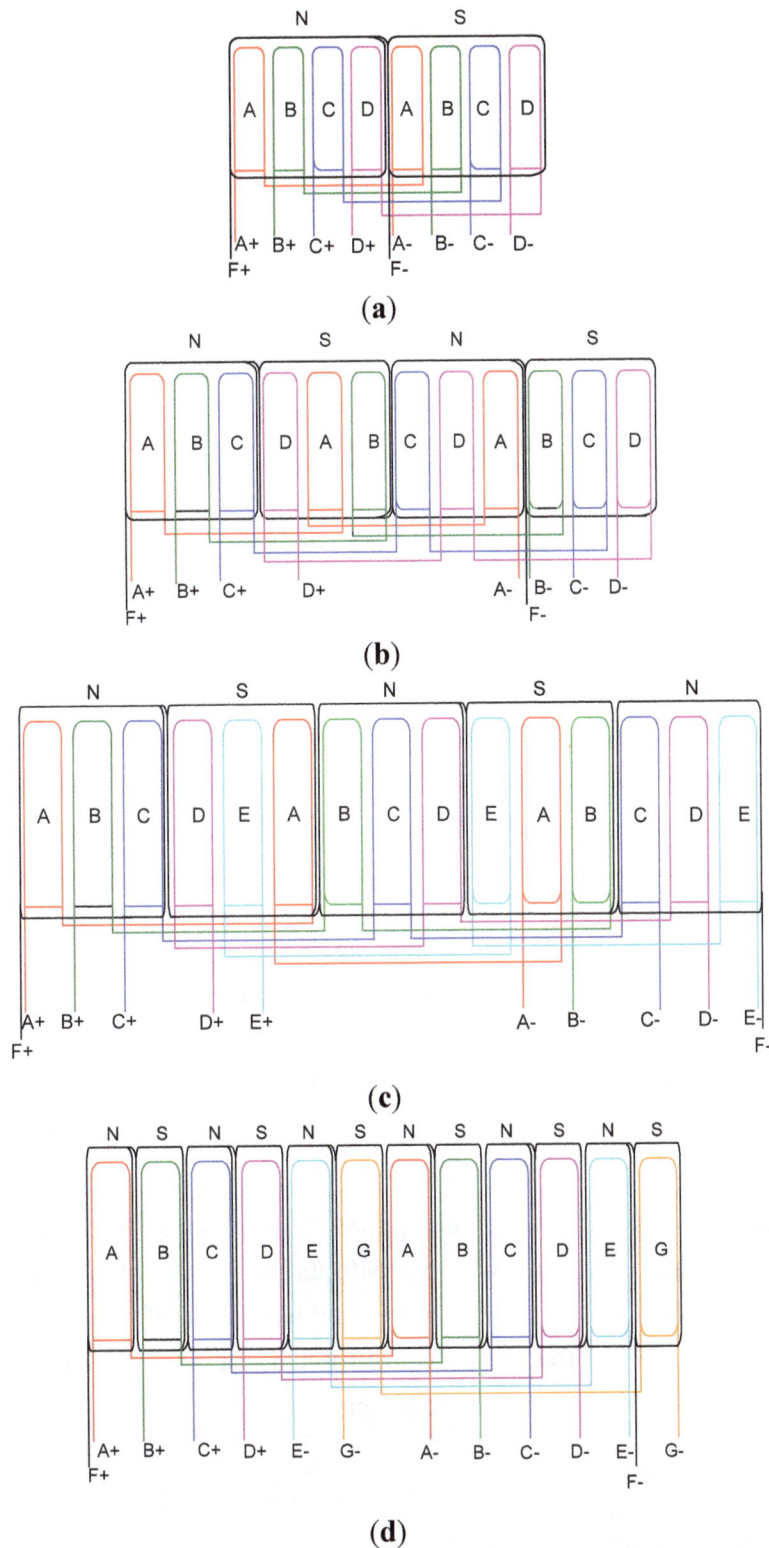

Figure 6. The connected coils of the multiphase WFDSM. (**a**) Traditional 8/6-pole four-phase; (**b**) new 12/9-pole four-phase; (**c**) 30/24-pole five-phase (half of the poles); (**d**) 12/10-pole six-phase.

The new 12/9-pole WFDSM has four field coils and twelve phase coils, which can be divided into four phases. As shown in Figure 6b, each field coil is wound around three stator poles, and the two neighboring field coils are in the opposite direction. Every phase has three coils, one is in the middle of the two field coil slots, and the other two phase coils are nearby the field coil slots. Therefore, the total reluctances of the four phases are equal. With the above analysis, the preferred configuration of the four-phase WFDSM with symmetry phases is with a 12/9-pole structure.

Similarly, we can draw the connected coils of the 30/24-pole five-phase WFDSM and the 12/10-pole six-phase WFDSM according to Table 2. Therefore, the elementary machine of the five-phase WFDSM with symmetry phases has a 30/24-pole structure.

In the six-phase WFDSM, if we wind the field coils around two or more stator poles, the six-phase WFDSM will still has the serious drawback of asymmetric phases, which will be verified by the simulation results in the next section Therefore, each stator pole of the 12/10-pole six-phase WFDSM should have a field coil if we want to get symmetrical phases.

3.2. Simulation Results

To compare the above multiphase WFDSMs, the simulation models of the traditional 8/6-pole and new 12/9-pole four-phase WFDSM were established. Figure 7a shows the flux of the 12/9-pole machine. Figure 7b shows the inductance between phase winding and field winding of the traditional 8/6-pole four-phase WFDSM. This verifies the formula of Equation (17). Figure 7c shows the same inductance of the new 12/9-pole four-phase WFDSG. It is shown that the amplitudes of the inductances have the relationship with:

$$\max(L_{af}) = \max(L_{df}) = \max(L_{bf}) = \max(L_{cf}) \tag{17}$$

The 2D-FEA results agree well with the theoretical analysis results. Figures 7d,e show the waveforms of the back EMF of the four-phase WFDSMs. The back EMFs of the traditional 8/6-pole WFDSM have slight difference in the amplitude and the shape of the waveform.

As the preferred five-phase WFDSM has a 30/24-pole structure, $j = 3$; its flux distribution is shown in Figure 8a. Figure 8b shows the inductances between the phase windings and the field winding of the traditional 10/8-pole five-phase WFDSM. The same inductance of the new 30/24-pole four-phase WFDSG is shown in Figure 8c. It shows that the traditional five-phase WFDSG has the disadvantage of asymmetric phases. The new WFDSG with its field coils wound around three poles can solve this problem. The theoretical analysis results are verified with these 2D-FEA results.

Figures 8d,e show the waveforms of the back EMF of the five-phase WFDSMs. It can be calculated that the back EMFs of different phases of the traditional 10/8-pole WFDSM have a difference of about 10.9%.

From the above analysis, we know that the traditional six-phase WFDSM with $j = 6$ has the phase asymmetry problem because the field coils is wound around six stator poles. We have set up the simulation models of the six-phase WFDSMs with different j. But even if we let $j = 5, 4$ or 3, there will be phase asymmetry problem.

If $j = 2$, the 12/10-pole six-phase WFDSM has a field coil wound around every two stator poles, and it also has the problem of asymmetric phases. As shown in Figure 9a, when the rotor pole is sliding to the pole of phase B, the back EMF of the phase B will be less than phase A because the pole of phase A has a lot of flux at this time. When the rotor pole is sliding to the pole of phase A, the back EMF of the phase A will be larger than phase B because the pole of phase B has no flux at this time. This is verified by the Figure 9d, which shows that the amplitudes of u_a and u_b are not equal. Therefore, the phases are still asymmetric even with $j = 2$.

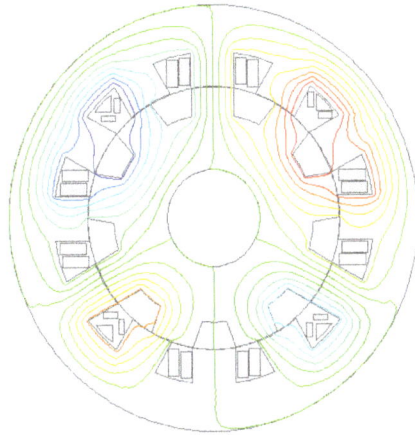

(a)

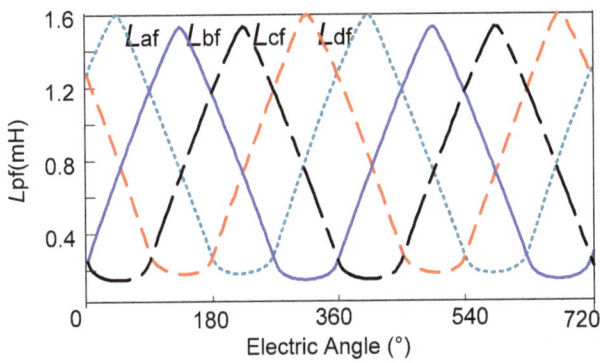

(b)

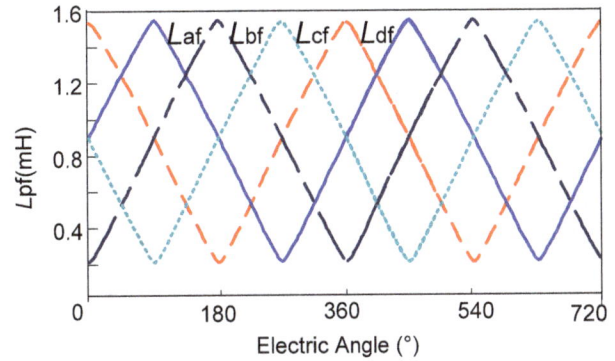

(c)

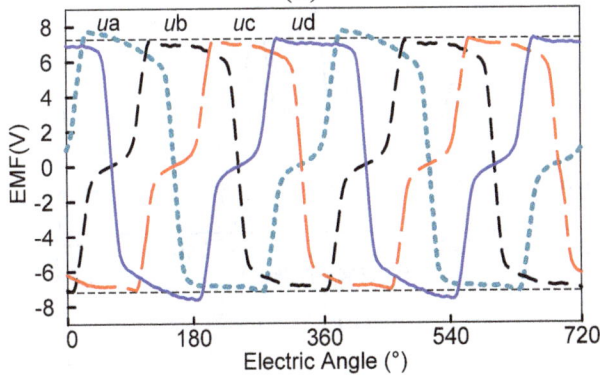

(d)

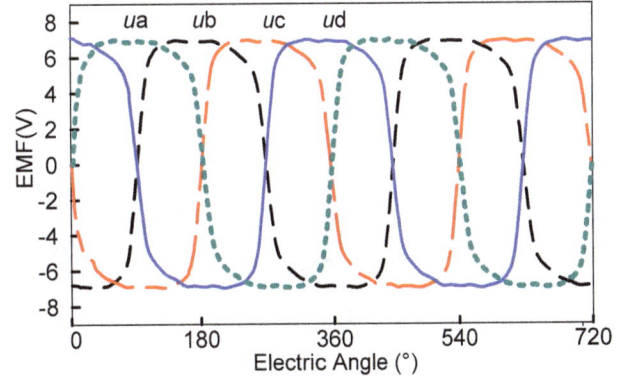

(e)

Figure 7. Comparison of the traditional and the new four-phase WFDSM. (**a**) The flux of the 12/9-pole WFDSM; (**b**) L_{pf} of the traditional 8/6-pole WFDSM; (**c**) L_{pf} of the new 12/9-pole WFDSM; (**d**) Back EMF of the 8/6-pole WFDSM; (**e**) Back EMF of the 12/9-pole WFDSM.

(a)

(b)

(c)

(d)

(e)

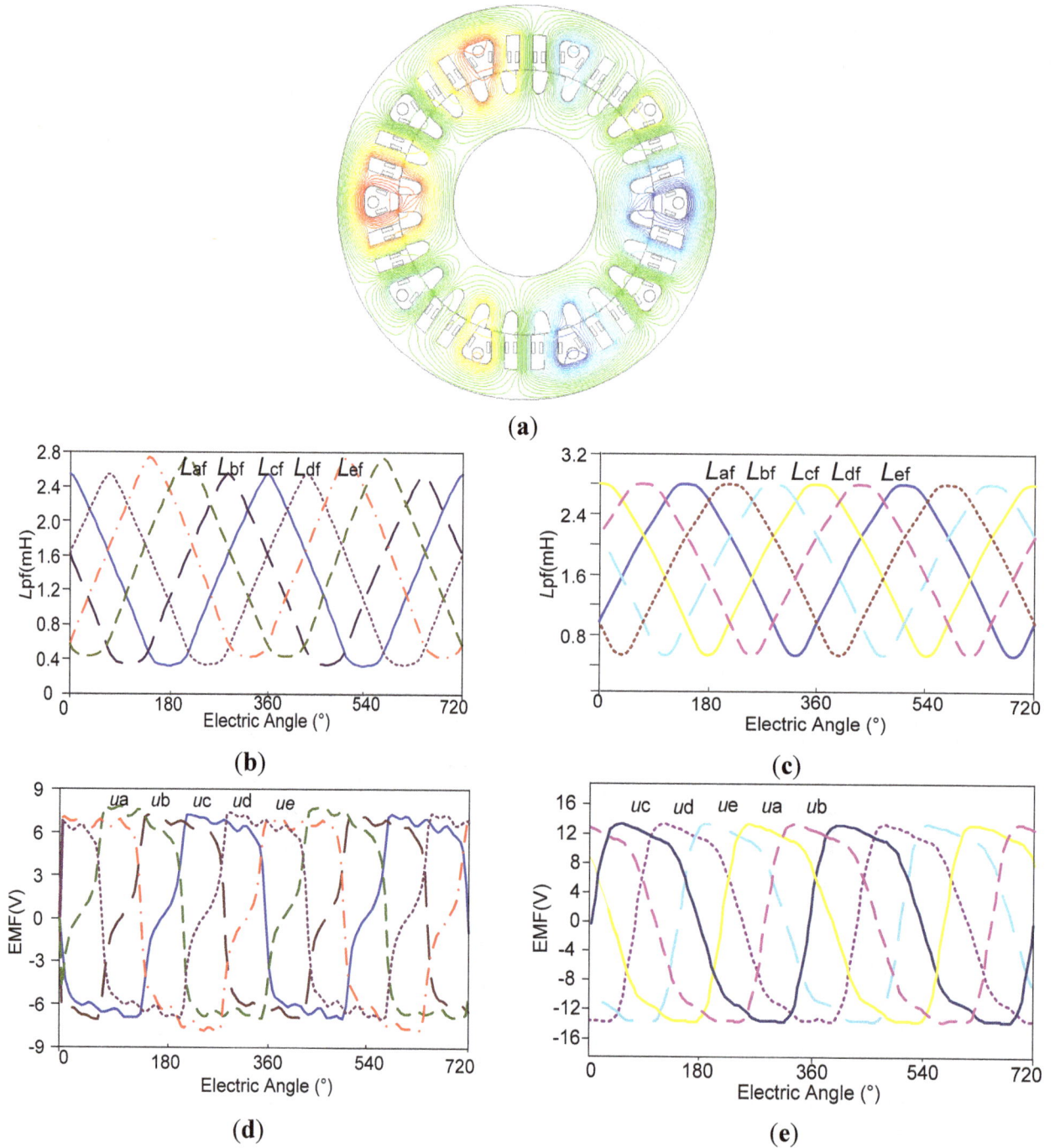

Figure 8. Comparison of the traditional and the new five-phase WFDSM. (**a**) The flux of the 30/24-pole WFDSM; (**b**) L_{pf} of the traditional 10/8-pole WFDSM; (**c**) L_{pf} of the new 30/24-pole WFDSM; (**d**) Back EMF of the 10/8-pole WFDSM; (**e**) Back EMF of the 30/24-pole WFDSM.

The flux of the six-phase WFDSM with $j = 1$ is shown in Figure 9b. This new WFDSG has its field coils wound around every stator pole, and solves the problem of phase asymmetry. The 2D-FEA results in Figure 9f verified this theoretical analysis. However, this new machine has a drawback of large copper loss, because there are field coils in every slot, and the resistance and the weight of the field winding will be increased.

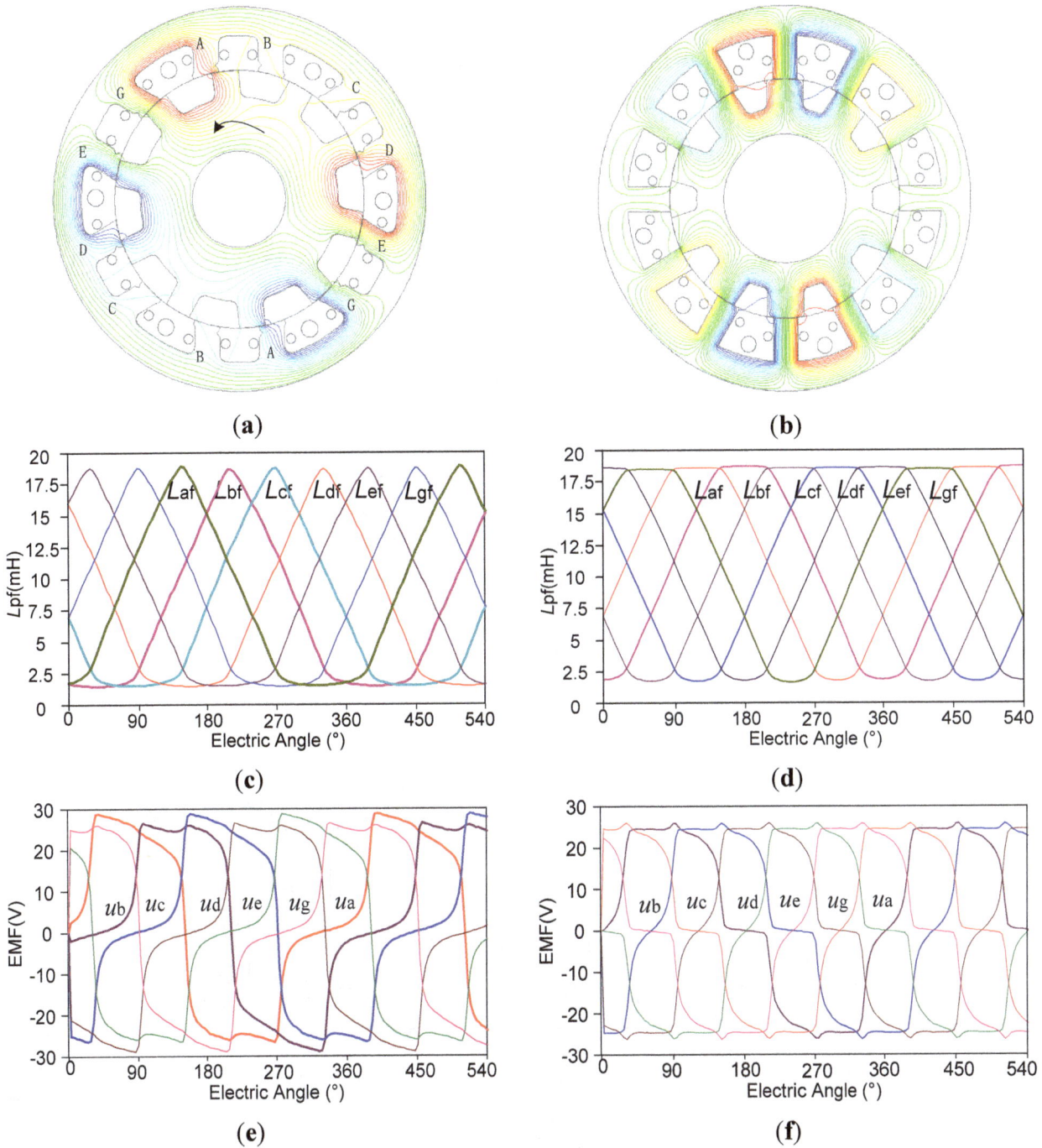

Figure 9. Comparison of the new 12/10-pole six-phase WFDSM. (**a**) The flux of the WFDSM with $j = 2$; (**b**) The flux of the WFDSM with $j = 2$. (**c**) L_{pf} of the WFDSM with $j = 2$; (**d**) L_{pf} of the WFDSM with $j = 1$; (**e**) Back EMF of the WFDSM with $j = 2$; (**f**) Back EMF of the WFDSM with $j = 1$.

4. Comparative Study of the Converter and Its Fault-Tolerant Performance

4.1. The Fault-Tolerant Converters

Multiphase machines can be divided into machines with prime number phases and machines with composite number phases. Because a composite number has at least one positive divisor other than 1 or the number itself, the machines with composite number phases can be divided into several channels which have little effect on each other [24]. For example, the four-phase WFDSM can be divided into two independent channels. However, the five-phase WFDSM has no positive divisor, and the machine may be susceptible to be suffer faults if the phase windings are connected together.

The fault-tolerant machine is usually equipped with a fault-tolerant converter. There are various types of fault-tolerant converters for different phase machines [25], and the four-phase converters will be discussed as an example in this paper.

The four-phase WFDSM is usually powered with a four-phase full bridge converter [26]. When there is a fault in one phase, the machine can keep on working because the fault is isolated from the other two phases, as shown in Figure 10a.

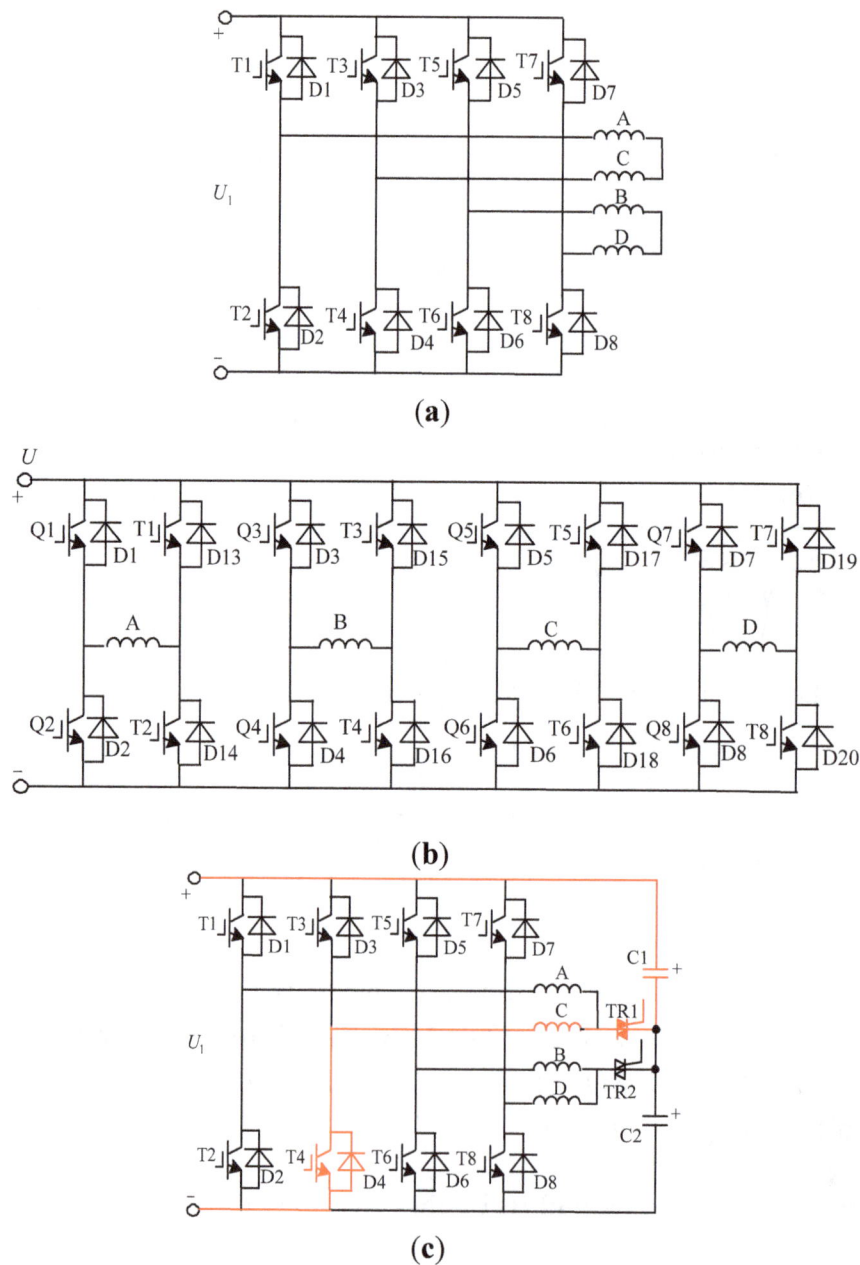

(a)

(b)

(c)

Figure 10. Comparison of the four-phase fault-tolerant converter. (**a**) The traditional four-phase converter; (**b**) The four-phase H bridges converter; (**c**) The half bridge four-phase converter.

The most excellent converter is the four-phase H bridges converter, as shown in Figure 10b, because all of the phases are isolated from each other. The machine can be designed as a modular machine, which offers potential fault-tolerant capability because the phase windings are isolated. When there is a fault in one phase, the other phases can keep on working without any infection from the fault phase. However, this converter design is not helpful to reduce the weight and the cost.

Figure 10c shows a half bridge four-phase converter. If phase A has an open circuit fault in this fault-tolerant converter, phase C of the machine can keep on operating with the help of split-phase capacitor C1. However, the five-phase WFDSM cannot be divided into two or three isolated channels, so it doesn't have such flexible fault-tolerant converters, and the faults of the machine may be susceptible to be infected if the phase windings are connected together.

As we can see from Figure 10, the switches increase with the phase number, and WFDSMs with more than six phases are not suitable because the weight and the cost are unacceptable. The WFDSMs with six phases or four phases can be divided into two isolated channels, which will improve the fault tolerance of the machine.

4.2. Fault-Tolerant Performance Comparison

The torque-angle characteristics of the multiphase WFDSMs are shown in Figure 11a. As we can see from the figure that the waveform of the 12/9-pole four-phase WFDSM is wider than the others, because the electrical angle of the phase voltage waveform of the four-phase WFDSM in Equation (8) is 180° and the phase voltage waveform electrical angle of the five-phase WFDSM is 144°, and the angle of the six-phase WFDSM is 120°.

There are four phases with mutative self-inductances at any time in all of the three WFDSMs, which can be described as $x = 4$ in Equation (8). If one phase of the four-phase WFDSM is isolated because of an open circuit or short circuit fault, the other three phases can keep on working. If the phase current maintains the same value because there is no fault, the fault-tolerant torque will be three-quarters of the normal torque. This is verified with Figure 11b.

To compare the fault-tolerant performance, the torque waveform with one phase open of the five-phase and the six-phase WFDSMs are shown in Figure 11b too. If the five steps in one period are named as five beats, then in the five beats of the five-phase WFDSM, there is a beat where the fault phase has no current. At this time, the output torque is equal to the normal torque. In the other beats, the fault-tolerant torque will be three-quarters of the normal torque too. The same result can be obtained with the six-phase WFDSM. This phenomenon is shown in Figure 11b. Because the WFDSMs are brushless DC machines, and their torque ripples are relatively large, which is mainly caused by the commutation torque ripple in the four-phase WFDSM. It should be noted that the commutation torque ripple of the WFDSM can be reduced by the optimal control method, which is investigated in [27]. Besides the commutation torque ripples, the five-phase and the six-phase WFDSMs also have the torque ripples caused by the differences between fault-tolerant beats and normal beats.

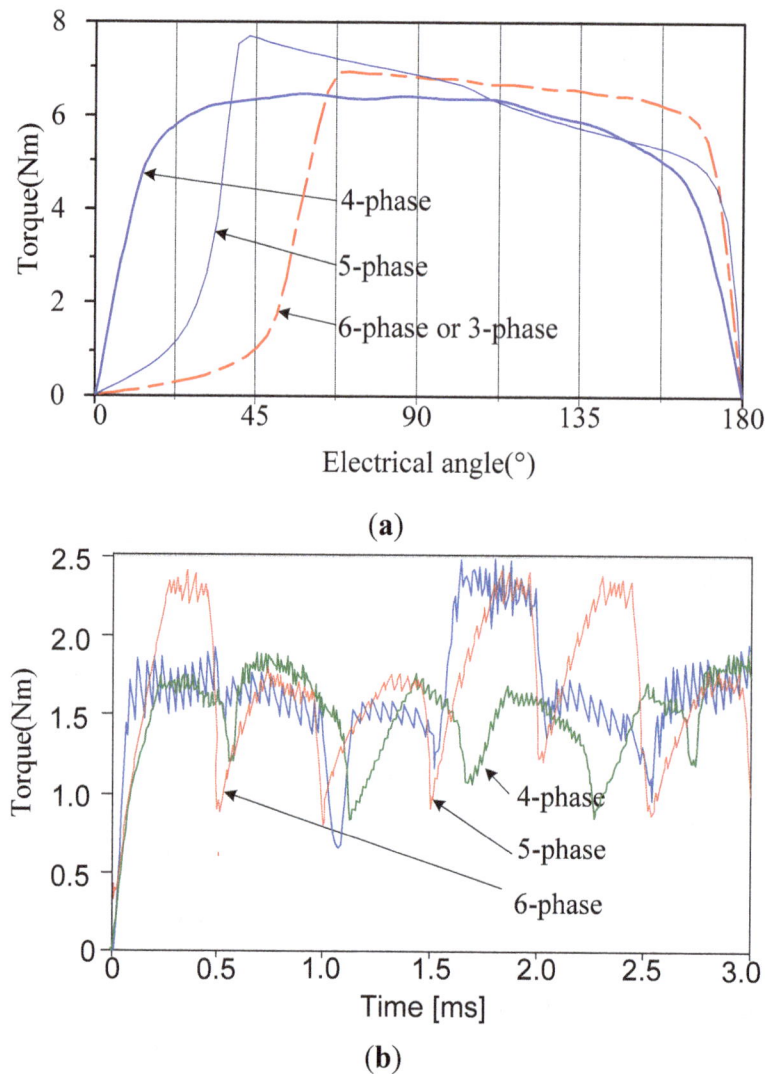

Figure 11. The torque of the multiphase WFDSMs. (**a**) The torque-angle characteristics; (**b**) The torque with one phase isolated in an H bridge converter.

What's more, as a brushless DC machine, the WFDSM is usually equipped with Hall sensors. Therefore, the five-phase WFDSM needs five Hall sensors, and the six-phase WFDSM needs only three Hall sensors, because phase A and phase D of the six-phase WFDSM have adverse EMF. The accuracy of the sensor should increase with the rotor pole number. Because the rotor pole number of the five-phase WFDSM with symmetrical phases is larger than the others, therefore, it is not an optimal solution for a fault-tolerant machine.

5. Conclusions

The four-phase WFDSM with $\alpha_s = 0.667$ has four changing inductance phases at any time, and its self-inductance of the field winding L_f will be a constant, which will not produce cogging torque. If it is used as a fault-tolerant machine, it has fewer switches in the converters than five-phase or six-phase machines.

The elementary machine of the five-phase WFDSM with symmetric phases has a 30/24-pole structure. It does not have cogging torque because the phase number with mutative inductance at any time is four. Different from the other five-phase machines with few poles, this five-phase WFDSM needs an expensive incremental encoder to provide sufficient position accuracy for the commutation control, but it can be a high performance fault-tolerant generator because it doesn't need position sensors.

If we wind the field coils around two or more stator poles, the six-phase WFDSM will have the serious drawback of asymmetric phases. The 12/10-pole six-phase WFDSM which has a field coil around each stator pole has symmetric phases, although it improves the copper loss of the field windings. Like the four-phase WFDSM, the six-phase WFDSM can be divided into two isolated channels, which will improve the fault tolerance of the machine. Because the switches of the converter and the pole numbers increase with the phase number, WFDSMs with more than six phases are not suitable because the weight and the cost of the converter are unacceptable, as well as the pole numbers.

Acknowledgments

This work was supported and funded by the National Natural Science Foundation of China (51477075), the Shandong Provincial Natural Science Foundation (ZR2014JL035), and the Fundamental Research Funds for the Central Universities (NP2015205).

Author Contributions

Bo Zhou conceived the idea of this paper, provide guidance and supervision; Li-Wei Shi implemented the research, performed the analysis and wrote the paper. All authors have contributed significantly to this work.

Conflicts of Interest

The authors declare no conflict of interest.

References

1. Villani, M.; Tursini, M.; Fabri, G.; Castellini, L. Electromechanical actuator for helicopter rotor damper application. *IEEE Trans. Ind. Appl.* **2014**, *2*, 1007–1014.

2. Boglietti, A.; Cavagnino, A.; Tenconi, A.; Vaschetto, S. The safety critical electric machines and drives in the more electric aircraft: A survey. In Proceedings of the 35th Annual Conference of IEEE Industrial Electronics, Porto, Portugal, 3–5 November 2009; pp. 2987–2995.

3. Bennett, J.W.; Atkinson, G.J.; Mecrow, B.C.; Atkinson, D.J. Fault-tolerant design considerations and control strategies for aerospace drives. *IEEE Trans. Ind. Electron.* **2012**, *5*, 2049–2058.

4. Garcia, A.; Cusido, J.; Rosero, J.A.; Ortega, J.A.; Romeral, L. Reliable electro-mechanical actuators in aircraft. *IEEE Aerosp. Electron. Syst. Mag.* **2008**, *8*, 19–25.

5. Villani, M.; Tursini, M.; Fabri, G.; Castellini, L. High reliability permanent magnet brushless motor drive for aircraft application. *IEEE Trans. Ind. Electron.* **2012**, *5*, 2703–2711.

6. Hao, L.; Du, H.Y.I.; Lin, H.; Namuduri, C. Design and analysis of PM fractional slot machine considering the fault operation. *IEEE Trans. Ind. Appl.* **2014**, *1*, 234–243.

7. Stewart, P.; Kadirkamanathan, V. Commutation of permanent-magnet synchronous AC motors for military and traction applications. *IEEE Trans. Ind. Electron.* **2003**, *3*, 629–630.

8. Du, Q.; Lu, E.; Shi, X. Design and analysis of five-phase tangent magnetic field permanent magnet generator for electric vehicle. *Int. J. Electric Hybrid Veh.* **2012**, *4*, 378–389.

9. Rottach, M.; Gerada, C.; Wheeler, P.W. Design optimisation of a fault-tolerant pm motor drive for an aerospace actuation application. In Proceedings of the 7th IET International Conference on Power Electronics, Machines and Drives, Manchester, UK, 8–10 April 2014; pp. 1–6.

10. Cossar, C.; Kelly, L.; Miller, T.J.E.; Whitley, C.; Maxwell, C.; Moorhouse, D. The design of a switched reluctance drive for aircraft flight control surface actuation. *IEE Colloq. Electr. Mach. Syst. More Electric Aircr.* **1999**, *11*, 1–8.

11. Hennen, M.D.; Hennen, M.D.; Heyers, C.; Brauer, H.J.; De Doncker, R.W. Development and control of an integrated and distributed inverter for a fault tolerant five-phase switched reluctance traction drive. *IEEE Trans. Ind. Electron.* **2012**, *2*, 547–554.

12. Gong, Y.; Chau, K.T.; Jiang, J.Z.; Yu, C.; Li, W. Design of doubly salient permanent magnet motors with minimum torque ripple. *IEEE Trans. Magn.* **2009**, *10*, 4704–4707.

13. Zhang, Z.; Yan, Y.; Tao, Y. A new topology of low speed doubly salient brushless DC generator for wind power generation. *IEEE Trans. Magn.* **2012**, *3*, 1227–1233.

14. Chen, Z.; Wang, H.; Yan, Y. A doubly salient starter-generator with two-section twisted-rotor structure for potential future aerospace application. *IEEE Trans. Ind. Electron.* **2012**, *9*, 3588–3595.

15. Liu, C.; Chau, K.T.; Zhong, J.; Li, J. Design and analysis of a HTS brushless doubly-fed doubly-salient machine. *IEEE Trans. Appl. Supercond.* **2011**, *3*, 1119–1122.

16. Liu, X.; Zhu, Z.Q. Electromagnetic performance of novel variable flux reluctance machines with DC-field coil in stator. *IEEE Trans. Magn.* **2013**, *6*, 3020–3028.

17. Yu, L.; Zhang, Z.; Chen, Z.H. Analysis and verification of the doubly salient brushless DC generator for automobile auxiliary power unit application. *IEEE Trans. Ind. Electron.* **2014**, *12*, 6655–6663.

18. Sui, Y.; Zheng, P.; Wu, F.; Yu, B.; Wang, P.F.; Zhang, J.W. Research on a 20-Slot/22-Pole five-phase fault-tolerant PMSM used for four-wheel-drive electric vehicles. *Energies* **2014**, *7*, 1265–1287.

19. Cheng, M.; Hua, W.; Zhang, J.; Zhao, W. Overview of stator-permanent magnet brushless machines. *IEEE Trans. Ind. Electron.* **2011**, *11*, 5087–5101.

20. Zhao, Y.; Wang, H.; Zhao, X.; Xiao, L. Characteristics analysis of five-phase fault-tolerant doubly salient electro-magnetic generators. In Proceedings of the IECON 2013—The 39th Annual Conference of the IEEE, Vienna, Austria, 10–13 November 2013; pp. 2668–2673.

21. Liu, X.; Zhu, Z.Q. Stator/Rotor pole combinations and winding configurations of variable flux reluctance machines. *IEEE Trans. Ind. Appl.* **2014**, *6*, 3675–3684.

22. Shi, J.T.; Liu, X.; Wu, D.; Zhu, Z.Q. Influence of stator and rotor pole arcs on electromagnetic torque of variable flux reluctance machines. *IEEE Trans. Magn.* **2014**, *11*, doi:10.1109/TMAG.2014.2330363.

23. Gaussens, B.; Hoang, E.; Barrière, O.D.; Saint-Michel, J.; Lecrivain, M.; Gabsi, M. Analytical approach for air-gap modeling of field-excited flux-switching machine: No-load operation. *IEEE Trans. Magn.* **2012**, *9*, 2505–2517.

24. Hill, C.I.; Zanchetta, P.; Bozhko, S.V. Accelerated electromechanical modeling of a distributed internal combustion engine generator unit. *Energies* **2012**, *5*, 2232–2247.

25. Bojoi, R.; Neacsu, M.G.; Tenconi, A. Analysis and survey of multi-phase power electronic converter topologies for the more electric aircraft applications. In Proceedings of the 2012 International Symposium on Power Electronics, Electrical Drives, Automation and Motion (SPEEDAM), Sorrento, Italy, 20–22 June 2012; pp. 440–445.

26. Chen, Z.; Chen, R.; Chen, Z. A fault-tolerant parallel structure of single-phase full-bridge rectifiers for a wound-field doubly salient generator. *IEEE Trans. Ind. Electron.* **2013**, *8*, 2988–2995.

27. Qin, H.; Wen, J.; Zhou, B.; Xue, H.H. Considerations of harmonic and torque ripple in a large power doubly salient electro-magnet motor drive. In Proceedings of the 2012 Asia-Pacific Symposium on Electromagnetic Compatibility (APEMC), Singapore, 22–24 May 2012; pp. 649–652.

9

Ash Content and Calorific Energy of Corn Stover Components in Eastern Canada

Pierre-Luc Lizotte [1], Philippe Savoie [1,2,*] and Alain De Champlain [3]

[1] Département des sols et de génie agroalimentaire, Université Laval, 2425 rue de l'Agriculture, Québec City, QC G1V 0A6, Canada; E-Mail: pierre-luc.lizotte@mail.mcgill.ca

[2] Agriculture and Agri-Food Canada, 2560 Hochelaga Blvd., Québec City, QC G1V 2J3, Canada

[3] Département de génie mécanique, Université Laval, 1065 avenue de la Médecine, Québec City, QC G1V 0A6, Canada; E-Mail: alain.dechamplain@gmc.ulaval.ca

* Author to whom correspondence should be addressed; E-Mail: philippe.savoie@fsaa.ulaval.ca

Academic Editor: Arthur J. Ragauskas

Abstract: Corn stover is an abundant agricultural residue that could be used on the farm for heating and crop drying. Ash content and calorific energy of corn grain and six stover components were measured from standing plants during the grain maturing period, between mid-September and mid-November. Ash of stover in standing corn averaged 4.8% in a cool crop heat unit zone (2300–2500 crop heat units (CHU)) and 7.3% in a warmer zone (2900–3100 CHU). The corn cob had the lowest ash content (average of 2.2%) while leaves had the highest content (from 7.7% to 12.6%). In the fall, ash content of mowed and raked stover varied between 5.5% and 11.7%. In the following spring, ash content of stover mowed, raked and baled in May averaged 3.6%. The cob and stalk located below the first ear contained the highest calorific energy with 17.72 MJ·kg^{-1}. Leaves and grain had the lowest energy with an average of 16.99 MJ·kg^{-1}. The stover heat of combustion was estimated at 17.47 MJ·kg^{-1} in the cool zone and 17.26 MJ·kg^{-1} in the warm zone. Based on presented results, a partial "cob and husk" harvest system would collect less energy per unit area than total stover harvest (44 *vs.* 156 GJ·ha^{-1}) and less biomass (2.51 *vs.* 9.13 t·dry matter (DM)·ha^{-1}) but the fuel quality would be considerably higher with a low ash-to-energy ratio (1.45 *vs.* 4.27 g·MJ^{-1}).

Keywords: corn stover; ash; calorific energy; corn components; cob; harvest

1. Introduction

In Eastern Canada, considerable energy is required on farms from October to April because of the prevailing cold weather. Heating is needed for several agricultural activities, notably greenhouses, various livestock buildings, grain drying and feed pelleting. At the present time, fossil fuels remain the most important source of energy on Canadian farms. Because crop residues are abundant and conveniently located in rural areas, they could become a sustainable source of thermal energy for agriculture. Such biomasses could reduce dependence to fossil fuels and reduce the global carbon footprint. For example in Québec, corn stover left in the field after grain harvest represents an estimated yield of 8.6 t·ha^{-1} and is available from an average area of 380,000 ha of grain-corn cultivated each year [1].

From data collected in the United States, corn stover could profitably be used as an energy source to dry corn grain on the farm [2]. This study used a calorific value of 16.5 MJ·kg^{-1} for corn stover but other researchers found higher energy content. Helsel and Wedin [3] recorded calorific values for corn stover between 16.4 and 17.4 MJ·kg^{-1}. They found that corn cob had the highest energy content among 12 different types of biomass. They also observed that calorific energy of corn varied on a yearly basis. Pordesimo *et al.* [4] also observed energy content ranging between 16.7 and 20.9 MJ·kg^{-1} during the grain harvest period. In Eastern Canada, small corn stover bales with an average of 18.57 MJ·kg^{-1} energy and 5.88% ash were used in a small furnace but resulted in incomplete combustion because of the high ash content [5].

The combustion of high ash biomass generally leads to solid agglomerates, greater emission of fumes, and accelerated metal wastage due to gas-side corrosion [6,7]. Handling large quantities of ash may also be cumbersome and costly. Ash is generally affected by hybrid variety, soil type, fertilization practices, and maturity [8]. Demirbas [9] reported that corn stover had a high natural ash content (5.1%) compared to wood (0.5% to 1.7%). However, most harvested biomasses have a higher ash level than their natural content due to soil contamination [10]. Ash content of corn stover in round bales was measured as high as 23.0% [11]. A low ash-to-energy ratio is therefore desirable and may be improved by appropriate handling of biomass in the field.

In Eastern Canada, field dry down of corn stover is considerably limited during the fall because of the prevailing cold and wet climate. Field conditions soon after grain harvest are generally inadequate for harvesting dry stover (<15% moisture content (MC)). However, leaving corn stover on the ground throughout the winter and harvesting it in the spring offer the possibility of collecting a very dry stover (<10% MC) [12,13]. Moreover, spring harvest provides extended soil protection against erosion by leaving a crop residue cover from November to April. However, less stover yield is available in spring compared to the previous fall because of natural degradation (average 21% less yield according to [13]). For combustion purposes, the winterization of corn stover may be beneficial by reducing the concentration of some minerals through leaching [14]. Reduction in minerals by delayed harvest was also reported for other energy crops such as miscanthus and switchgrass [15,16]. Nevertheless, there are very few studies on the effects of delayed harvest on fuel quality of corn stover. Therefore, the objectives of this study were (1) to measure ash content and calorific energy of corn stover during an extended period in the fall; (2) to evaluate the ash content of stover baled in the spring; and (3) to simulate various biomass harvest scenarios for combustion.

2. Materials and Methods

2.1. Harvest Sites and Sample Preparation

Corn stover was obtained from two sources: (1) standing crop prior to grain harvest; (2) field stover chopped after grain harvest and collected from windrows or from bales. Standing corn plants were sampled weekly in experimental plots during the grain maturing period in two climatic zones from early September to mid-November 2008. Hybrids Elite 46T07 and Elite 30A27 were grown in Saint-Augustin-de-Desmaures, QC, Canada (46.732° N, 71.517° W) in a relatively cool zone of 2300 to 2500 crop heat units (CHU). Hybrids Elite 46T07 and Elite 25T19 were grown in Saint-Rosalie, QC, Canada; (45.606° N, 72.914° W) in a warmer zone of 2900 to 3100 CHU. Each week, five corn plants from each hybrid and site were partitioned into grain and six stover components (cob, husk, upper leaves, lower leaves, upper stalk, lower stalk). Total biomass and moisture content have been reported in [1]. The line of separation of lower leaves and upper leaves was the first ear, as well as for lower and upper stalks. Each corn component was ground with a laboratory mill (model ED-5, Thomas Scientific, Swedesboro, NJ, USA) using a 1-mm screen.

In large commercial corn fields (average 2-ha plots), stover was mechanically handled after grain harvest at two sites (Table 1). Site 1 was located in La Présentation, QC, Canada (45.615° N, 73.070° W) in a 2900–3100 CHU zone where grain was harvested on 1 November 2008. Site 2 was located 44 km South-West from the first site in Saint-Philippe-de-Laprairie, QC, Canada (45.318° N, 73.451° W, 2900–3100 CHU zone) where grain was harvested on 12 November 2008. Besides two plots windrowed in fall 2008, three other plots, one at site 1 and two at site 2, were mowed and raked in spring 2009. All five plots were baled in spring 2009 at dates and with equipment indicated in Table 1. Ten samples were taken from large square bales and ten other samples from large round bales, while 20 samples were taken from small square bales for calorific value and ash content measurements per plot. Immediately after baling, stover samples were taken off the bale surface rather than by coring because of the difficulty of tube sampling through the very dry and brittle biomass.

Windrows were also sampled weekly in the fall until the first snowfall and once in the spring before baling. At site 1, windrows were actually sampled from 5 November to 3 December (five weekly samples) and on 30 April. At site 2, windrows were sampled from 19 November to 3 December (three weekly samples) and on 24 May. Grab samples were taken at 51 mm from the top and at 51 mm from the bottom of the windrow to avoid soil contamination.

Table 1. Sequence of operations for each plot, dates and equipment to collect stover after grain harvest in fall 2008. Stover windrows were baled in spring 2009 [12].

Date of operation	Site 1		Site 2		
	Fall Plot	Spring Plot	Fall Plot	Spring Plot	Spring Plot
	(1.83 ha)	(3.08 ha)	(2.16 ha)	(2.68 ha)	(1.68 ha)
5 Nov. 2008	Mowing [1]	-	-	-	-
	Raking [2]	-	-	-	-
19 Nov. 2008	-	-	Mowing [3]	-	-
	-	-	Raking [2]	-	-
26 Apr. 2009	Raking [2]	Mowing [1]	-	-	-
	-	Raking [2]	-	-	-
29 Apr. 2009	Raking[2]	Raking [2]	-	-	-
30 Apr. 2009	Large rect. baling [5]	Large rect. baling [5]	-	-	-
24 May 2009	-	-	Raking [2]	-	-
25 May 2009	-		Round baling [6]	Shred-windrow [4]	Shred-windrow [4]
	-	-	-	Round baling [6]	Small rect. baling [7]

Notes: Equipment used: [1] Disc mower, New Idea model 5208, 2.4 m wide (Duluth, GA, USA); [2] Parallel bar rake, New Holland model 5208, four windrow wide or 9.8 m (Burr Ridge, IL, USA); [3] Disc mower, Taarup model 2424, 2.4 m wide (Kvernaland, Norway); [4] Flail shredder-windrower, Hiniker model 5610, 4.5 m wide (Mankato, MN, USA); [5] Large rectangular baler, Case IH model LBX332, typical bale size 1.83 m × 0.90 m × 0.81 m (Burr Ridge, IL, USA); [6] Large round baler, John Deere model 458, typical bale size 1.22 m diameter × 1.22 m width (Moline, IL, USA); [7] Small rectangular baler, New Holland model 315, typical bale size 0.35 m × 0.45 m × 0.81 m (New Holland, PA, USA).

2.2. Combustion Tests

The calorific energy of the seven components of standing crops was measured for samples taken during week 3 (17 September 2008), week 6, week 9, and week 12 (19 November 2008) using an 1108P bomb inserted in a 6100 calorimeter (Parr Instrument Co., Moline, IL, USA) calibrated for uniform measurements. Three replicates were made from about 1 g of kernel and cob, and 0.7 g of husk, leaf and stalk material, transformed into pellets and weighed with a precision scale of 0.0001 g. Pellets were inserted into a bomb pressurized to 3.1 MPa with 99.5% oxygen purity. Calibration tests were conducted after every 40 runs using benzoic acid capsules of 26.4 MJ·kg^{-1} (6 318.4 cal·g^{-1}); the average deviation was 0.29 MJ·kg^{-1}.

2.3. Ash Measurements

Ash content of stover components of standing crops was measured with a TGA701 Leco thermogravimetric analyzer (St. Joseph, MI, USA). Ash was expressed as the mass percent of residue remaining after 3 h of dry oxidation at 525 °C based on ASTM E1755-01 Standard [17]. Because of limited instrument availability, only 54 samples were evaluated for ash. Samples were selected among four weeks (3, 6, 9, and 12), two hybrids, two sites and six components to allow one-factor analyses of ash content. Ash content was also measured for raked and baled corn stover representing a composite mix of components. Samples were taken from windrows or bales as described in the previous section. Samples were analyzed by the same procedure as for standing crops [17].

For both calorific and ash data, a repeated analysis of variance (ANOVA) was performed to measure the effect of time. A paired Student's t-test was used to conduct simple comparison while Tukey's Honestly Significant Difference (HSD) test was used for multiple pairwise comparisons. A significance level (α) of 0.05 was used and the R software was chosen for statistical computations [18].

2.4. Simulation of Stover Harvest

A simulation was done to estimate potential stover harvest and its quality in terms of ash content and calorific energy, especially in the context of on-farm use for animal bedding and small-scale heating which appear to be the main potential applications in Eastern Canada [5,13]. Four scenarios were considered. The first scenario assumes that only cobs are harvested behind a combine (e.g., H165 Cob harvester; Redekop, Saskatoon, SK, Canada). The second scenario collects cob and husk with a cob harvester and stover processor (e.g., Cob Collection Attachment; Hillco Technologies, Nezperce, ID, USA). The third scenario collects all stover components above the ear; this can be achieved by using a combine equipped with a forage harvester header and a stover processor [19]. The fourth scenario uses equipment similar to the third scenario; the whole stover is harvested by lowering the forage header closer to the ground. Simulation results compare total biomass, ash and energy for the four scenarios.

3. Results and Discussion

3.1. Ash Content of Standing Corn Plants

Ash content of standing corn stover was compared among hybrids, sites and components at four dates between September and November, just before grain harvest. The effect of time on stover ash content revealed a non-significant effect at a p-value of 0.503 (54 data points). According to a paired Student's t-test analysis, differences in ash between hybrids grown at a same site were non-significant at a probability of 0.108 (16 data points collected at the 2900–3100 CHU site). Values were therefore averaged over time and hybrids (Table 2). Corn grown in the high CHU zone had a significantly ($p < 0.01$; 26 data points) higher ash content (average 7.3%) than corn grown in the lower CHU zone (average 4.8%). However, statistical results did not show a significant difference ($p > 0.05$; 10 data points) between ash content of the 46T07 hybrid (a 2300 CHU hybrid) sampled in either the low CHU zone or the high CHU zone. The significant difference in ash content between the two CHU zones may be explained by a combination of factors such as microclimate, morphological differences between hybrids, soil type, and fertilization practices within each site.

For both CHU zones, the husk and cob fractions had the lowest ash content with average values ranging between 2.07% and 3.20% (Table 2). Leaves were the stover component with the highest ash content from 7.69% to 12.59%; there was no significant difference between upper and lower leaves. The ash content difference was also non-significant between lower and upper stalk fractions. Xiong *et al.* [20] reported corn stalks and leaves containing ash concentrations of 4.31% and 8.37%, respectively for hybrids grown in China. Those values are close to ash levels recorded at the 2300–2500 CHU site.

Table 2. Average ash content of corn stover components from standing crops sampled in two crop heat unit (CHU) zones during grain maturing period (from 17 September to 19 November 2008).

Corn Stover Components	2300–2500 CHU		2900–3100 CHU	
	Mass Ratio (%)	Ash Content (%) [a]	Mass Ratio (%)	Ash Content (%)
Cob	21.8	2.07 [a]	18.6	2.26 [a]
Husk	8.4	2.53 [a]	8.9	3.20 [a]
Lower leaves	12.8	7.69 [b,c]	15.0	12.59 [c]
Upper leaves	13.8	8.13 [c]	17.2	12.44 [c]
Lower stalk	32.8	4.62 [b]	28.8	6.59 [b]
Upper stalk	10.5	4.92 [b,c]	11.4	5.86 [b]
Tukey's HSD	-	2.68	-	1.92
Total weighted stover	100.0	4.80	100.0	7.31

Notes: Different letter superscripts (a, b, c) in the same column denote significant differences among stover components according to the Tukey's HSD test ($\alpha = 0.05$).

3.2. Ash Content of Windrowed and Baled Corn Stover

Ash content of raked corn stover at site 1 (La Présentation, QC, Canada) varied between 9.42% and 11.69% from 5 November (day of raking) to 3 December 2008 (Figure 1). There appears to be some initial soil contamination once the stover is on the ground. An important rainfall (30.4 mm) occurred on 15 November which could explain an ash content reduction due to leaching at site 1 (9.42% ash recorded on 19 November). However, wind or splashing due to light rain could have re-contaminated the windrows in subsequent weeks. At site 1, the effect of time was non-significant on ash content (p-value of 0.612). However, a significant time effect (p-value < 0.001) on ash content was observed at site 2 (Saint-Philippe-de-Laprairie). Indeed, ash increased from 5.49% on 19 November to 10.46% on 3 December, where wind and light rain could have caused soil contamination over time.

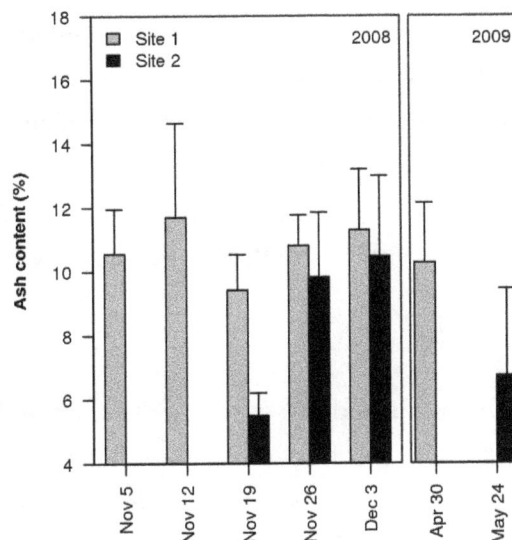

Figure 1. Ash content of windrowed corn stover left on the ground at two sites of similar climate (2900–3100 CHU) sampled in fall 2008 and in spring 2009. Error bars denote the standard deviation.

In spring 2009, average ash of windrowed stover was 10.27% and 6.73% at sites 1 and 2, respectively. For site 1, ash content of windrows sampled in the spring was not significantly different (p-value = 0.399) to the measurements made on 3 December 2008 (11.29%). At site 2, ash was significantly lower (p-value < 0.001) in the spring than on 3 December 2008 (10.46%). The high standard deviations of ash in windrowed stover (1.82%) reflect heterogeneity due to soil contamination and variability in sampling.

Table 3 shows ash of bales from two sites, mowed and raked either in fall or spring, with all bales harvested in spring. On site 1 (La Présentation), bale ash of spring-mowed stover was 5.21% and significantly lower (p-value = 0.022) than bale ash of fall-mowed stover (7.03%). On site 2 (Saint-Philippe-de-Laprairie), bale ash was also lower in the spring-mowed stover (3.59%) but not significantly different (p-value = 0.085) from fall-mowed stover (4.33%). For three plots out of four, the standard deviation of bale ash was lower than the average value of 1.82% for weekly sampled windrows (Figure 1).

Table 3. Ash content of corn stover bales at two sites for two mowing-raking periods (fall or spring). All windrows were baled in spring.

Site	Mowing and Raking Period	Bale Ash (%) Average *	Std Deviation
Site 1 (La Présentation)	Fall 2008	7.03 [b]	1.83
	Spring 2009	5.21 [a]	1.48
Site 2 (Saint-Philippe de Laprairie)	Fall 2008	4.33 [a]	1.18
	Spring 2009	3.59 [a]	0.52

Note: * Different letter superscripts (a, b) denote significant differences according to Student's t-test ($\alpha = 0.05$).

Bales mowed and raked in the spring had a lower ash content than bales mowed and raked in the previous fall. In China, stover ash of standing corn declined from 15% in late August to 6.4% in mid-March [21]. Baled stover with the lowest ash values (Site 2, Spring 2009) was actually handled with a single pass shredder-windrower while other fields were handled in two separate operations of mowing and raking. The combination of spring harvest and simultaneous mowing-conditioning appears to reduce ash content by winter leaching and minimal soil contamination until baling. The very low ash content (3.6%) of spring harvested stover with simultaneous shredding-windrowing also compares favorably to a single-pass grain and stover fall harvest system in Wisconsin where stover ash averaged 4.9% [22]. These latter researchers also tested multi-pass harvest of stover in the fall with flail-shredding, raking, and chopping. Stover ash collected with the multi-pass scenario averaged 9.8%.

3.3. Calorific Energy of Corn Components

Calorific energy was measured from standing corn plants sampled from 17 September to 19 November 2008, every three weeks. The cob fractions had the highest calorific energy with values ranging from 17.70 to 18.34 MJ·kg^{-1} (Figure 2). Meanwhile, leaves (averaged for lower and upper fractions) had the lowest energy from 16.60 to 17.04 MJ kg^{-1}. Cob results were similar to the 17.10 to 18.16 MJ·kg^{-1} range reported by [3]. For all corn components, the time effect was statistically non-significant although the probability level for cob was close to α with 0.068. In fact, cob energy increased by 4% from 29 October to 19 November but data were limited at the last date because only one site (2900–3100 CHU) was sampled on 19 November.

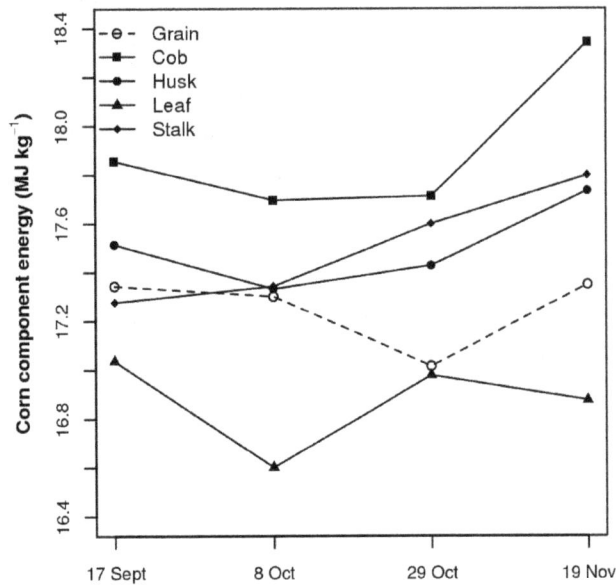

Figure 2. Corn component energy of standing crop as a function of sampling date, averaged for hybrids and sites.

Table 4 shows calorific values of corn stover components at two sites on a specific date (29 October). Over the two sites, the cob and upper-stalk had significantly higher calorific energies than the other corn components with values close to 17.70 MJ·kg^{-1}. For the 2300–2500 CHU zone, the component with the lowest energy was the grain with 16.98 MJ·kg^{-1}. The leaves located below the first ear had the lowest energy (16.98 MJ·kg^{-1}) for the 2900–3100 CHU zone. According to work by [20], corn stalk (18.92 MJ·kg^{-1}) had a greater energy than leaves (17.99 MJ·kg^{-1}). They reported energy values as high as 19.35 MJ·kg^{-1} for stalk sampled in China. On 29 October, the weighted average of stover gross energy was generally higher (17.47 MJ·kg^{-1}) for hybrids grown in the 2300–2500 CHU zone.

Table 4. Calorific energy of corn plant components sampled from the standing crop on 29 October 2008.

Corn Component	Calorific Energy (MJ·kg^{-1}) [a]		
	2300–2500 CHU	2900–3100 CHU	Overall
Grain	16.98 [a]	17.12 [b,c]	17.01 [a]
Cob	17.62 [b]	17.80 [c]	17.71 [b]
Husk	17.49 [a,b]	17.36 [c]	17.43 [a,b]
Lower-leaves	17.38 [ab]	16.18 [a]	16.92 [a]
Upper-leaves	17.57 [b]	16.35 [a,b]	17.04 [a]
Lower-stalk	17.50 [ab]	17.37 [c]	17.46 [a,b]
Upper-stalk	17.87 [b]	17.56 [c]	17.74 [b]
Tukey's HSD ($\alpha = 0.05$)	0.54	0.83	0.56
Stover	17.47	17.26	17.41

Notes: Different letter superscripts (a, b, c) in same column denote significant differences among stover components according to the Tukey's HSD test ($\alpha = 0.05$).

During the overall sampling period, the calorific energy of corn stover was significantly higher (*p*-value of 0.0011) for the 2300–2500 CHU zone with values ranging from 17.27 to 17.64 MJ·kg^{-1} (Figure 3). Those results corroborate with the low ash content of corn stover measured for the 2300–2500 CHU hybrids (Table 2). For the 2900–3100 CHU zone, stover energy was low in mid-September with values around 17.00 MJ·kg^{-1} and increased to 17.50 MJ·kg^{-1} toward the end of the sampling period. Energy of stover at the 2900–3100 CHU site ranged from 16.86 to 17.45 MJ·kg^{-1}. Those results were very close to the 16.42 to 17.41 MJ·kg^{-1} measured by [3] for corn stover sampled in 1974 and 1975. However, values reported by [4] for corn plants sampled from 9 August to 26 November 2001 in Tennessee were in the lower end of the 16.7 to 20.9 MJ·kg^{-1} range. The calorific energy of the 46T07 hybrid grown in the 2300–2500 CHU zone was higher than the same hybrid sampled in the 2900–3100 CHU zone. However this difference was not statistically significant with a resulting paired Student's *t*-test of 0.082.

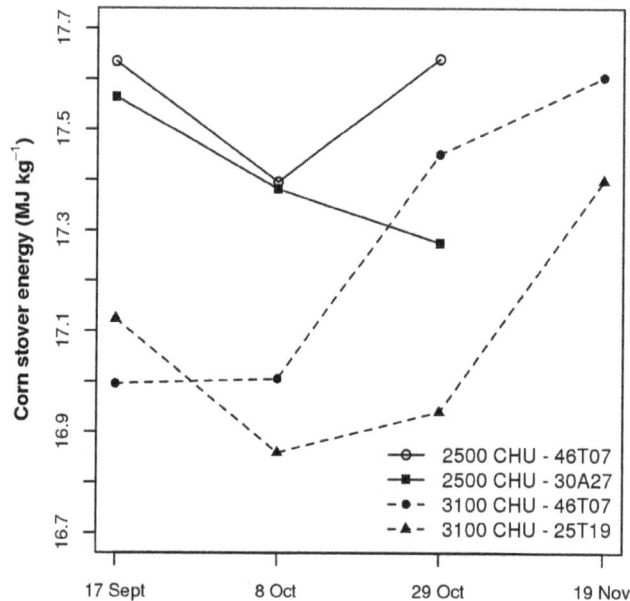

Figure 3. Calorific energy of corn stover from various standing hybrids in two crop heat unit (CHU) zones.

3.4. Harvesting Corn Stover for Combustion

Few studies have highlighted the possibility to enhance biomass quality by harvesting only specific corn stover components [19,23]. The simulation results in Table 5 are based on standing crop stover quality observed in two climatic zones described in the materials and methods section.

Corn cob was the stover component with the lowest ash content and the highest calorific energy. So all harvest scenarios collecting other components than cob had a greater ash content and a lower unit energy value. Cob harvest yielded 25 and 30 GJ·ha^{-1} in the 2300–2500 and 2900–3100 CHU zones, respectively. For this first scenario, a higher energy yield was calculated for the 2900–3100 CHU zone since the calorific energy and the cob biomass yield were greater than for the 2300–2500 CHU zone. The unit calorific energy of whole stover harvest was the lowest for both zones but the total energy yield was the highest with 116 and 156 GJ·ha^{-1} for the low and high CHU zones, respectively. Whole stover harvest also resulted in the highest ash yields considering ratios of ash-to-energy of 2.73 and 4.27 g·MJ^{-1} for low and high CHU zones. For all scenarios of the 2300–2500 CHU zone, ash-to-energy ratios were lower than those calculated for the 2900–3100 CHU zone. The largest increases of ash-to-energy ratio were observed when more fractions than cob and husk were harvested.

Table 5. Biomass, ash, and energy yield of corn stover for four harvest scenarios.

Harvest Scenario	Ash Content (%)	Calorific Energy (MJ·kg^{-1})	Ash to Energy Ratio (g·MJ^{-1})	Yield	
				Biomass (t·DM·ha^{-1})	Energy (GJ·ha^{-1})
2300 to 2500 Crop Heat Unit zone					
Cob	2.07	17.62	1.17	1.44	25
Cob + Husk	2.20	17.58	1.25	1.99	35
Stover above ear [a]	4.22	17.64	2.39	3.60	64
Whole stover	4.80	17.56	2.73	6.62	116
2900 to 3100 Crop Heat Unit zone					
Cob	2.26	17.80	1.27	1.70	30
Cob + Husk	2.56	17.66	1.45	2.51	44
Stover above ear [a]	6.26	17.24	3.63	5.13	88
Whole stover	7.31	17.12	4.27	9.13	156

Notes: [a] Stover above the ear includes cob, husk, upper-leaves and upper-stalk.

4. Conclusions

Ash content and calorific energy of corn stover were measured on standing corn plants sampled in the fall. Additional ash measurements were taken on corn stover from windrows left in the field up to six months after grain harvest and from bales harvested in the spring. Average value of ash in standing stover was 4.8% in a cool 2300–2500 crop heat unit (CHU) zone and 7.3% in a warmer 2900–3100 CHU zone. Ash content in standing plants did not change significantly between September and November. Higher ash levels were found in windrowed stover, in the range of 5.5% to 11.7% in November and December after grain harvest. Corn stover bales harvested in the following spring contained ash levels between 3.6% and 7.0%. The lowest stover ash content was observed in plots that were mowed and raked simultaneously in the spring, and then baled soon after. Leaving the standing corn stover throughout the winter was beneficial for mineral leaching and ash reduction.

The calorific energy of standing corn stover varied between 16.86 and 17.64 MJ·kg^{-1}. The effect of time between September and November was non-significant. However, stover energy was statistically higher for hybrids grown in the cooler, 2300–2500 CHU zone. The cob fraction had the highest calorific energy among corn components with values around 17.71 MJ·kg^{-1}. Four harvest scenarios were analyzed. The whole stover scenario resulted in the highest total energy yield of 116 and 156 GJ·ha^{-1} for the 2300–2500 and the 2900–3100 CHU zones, respectively. This scenario also yielded the highest ash-to-energy ratios (2.73 and 4.27 g of ash·MJ^{-1}, in the two zones respectively). The selective harvest of only cob as a solid fuel resulted in lower ash to energy ratios (1.17 and 1.27 g·MJ^{-1}). Partial harvest may be considered if a reduction of the ash-to-energy ratio is an important factor.

Acknowledgments

Authors wish to thank La Coop fédérée, Uniboard Canada and Agriculture and Agri-Food Canada for supporting a Matching Investment Initiative (MII) project related to corn stover feedstock. The authors also acknowledge the Natural Science and Engineering Research Council of Canada (NSERC) for continued support in training highly qualified personnel through its Discovery grant program. Assistance for data collection was provided by Marie-Pierre Fortier and laboratory work was conducted by Héloïse Bastien and Andrée-Dominique Baillargeon.

Author Contributions

Pierre-Luc Lizotte and Philippe Savoie designed the experiments. Pierre-Luc Lizotte performed the experiments, analyzed the data and drafted the paper under supervision of Philippe Savoie and Alain De Champlain. Pierre-Luc Lizotte, Philippe Savoie and Alain De Champlain discussed results, reviewed and finalized the paper.

Conflicts of Interest

The authors declare no conflict of interest.

References

1. Lizotte, P.-L.; Savoie, P.; Lefsrud, M.; Allard, G. Yield and moisture content of corn stover components in Québec, Canada. *Can. Biosyst. Eng. J.* **2014**, *56*, 8.1–8.9.

2. Bennett, A.S.; Bern, C.J.; Richard, T.L.; Anex, R.P. Corn grain drying using corn stover combustion and CHP systems. *Trans. ASABE* **2007**, *50*, 2161–2170.

3. Helsel, Z.R.; Wedin, W. Direct combustion energy from crops and crop residues produced in Iowa. *Energy Agric.* **1983**, *1*, 317–329.

4. Pordesimo, L.O.; Hames, B.R.; Sokhansanj, S.; Edens, W.C. Variation in corn stover composition and energy content with crop maturity. *Biomass Bioenergy* **2005**, *28*, 366–374.

5. Morissette, R.; Savoie, P.; Villeneuve, J. Combustion of Corn Stover Bales in a Small 146-kW Boiler. *Energies* **2011**, *4*, 1102–1111.

6. Johansson, L.S.; Tullin, C.; Sjövall, P. Particle emissions from biomass combustion in small combustors. *Biomass Bioenergy* **2003**, *25*, 435–446.

7. Loo, S.V.; Koppejan, J. *The Handbook of Biomass Combustion and Co-Firing*; Earthscan: London, UK, 2008.

8. Samson, R.; Mehdi, B. *Strategies to Reduce the Ash Content in Perennial Grasses*; Resource Efficient Agricultural Production-Canada: Sainte-Anne-de-Bellevue, QC, Canada, 1998.

9. Demirbas, A. Combustion characteristics of different biomass fuels. *Prog. Energy Combust. Matter* **2004**, *30*, 219–230.

10. Hoffman, P.C. *Ash Content of Forages*; University of Wisconsin-Extension: Marshfield, WI, USA, 2005.

11. Wright, C.L. Effect of harvest method on the nutrient composition of baled cornstalks. In *2005 South Dakota Beef Report*; South Dakota State University: Brookings, SD, USA, 2005; pp. 46–47.

12. Lizotte, P.-L.; Savoie, P. Spring harvest of corn stover. *Appl. Eng. Agric.* **2011**, *27*, 697–703.

13. Lizotte, P.-L.; Savoie, P. Spring harvest of corn stover for animal bedding with a self-loading wagon. *Appl. Eng. Agric.* **2013**, *29*, 25–31.

14. Schreiber, J.D. Nutrient leaching from corn residues under simulated rainfall. *J. Environ. Qual.* **1999**, *28*, 1864–1870.

15. Lewandowski, I.; Heinz, A. Delayed harvest of miscanthus—Influences on biomass quantity and quality and environmental impacts of energy production. *Eur. J. Agron.* **2003**, *19*, 45–63.

16. Adler, P.R.; Sanderson, M.A.; Boateng, A.A.; Weimer, P.J.; Jung, H.-J.G. Biomass yield and biofuel quality of switchgrass harvested in fall or spring. *Agron. J.* **2006**, *98*, 1518–1525.

17. *Standard Test Method for Ash in Biomass*; ASTM E17755-01(2007); ASTM International: West Conshohocken, PA, USA, 2007.

18. R Core Team. *R: A Language and Environment for Statistical Computing*; Version 3.1.1.; R Foundation for Statistical Computing: Vienna, Austria, 2014.

19. Shinners, K.J.; Adsit, G.S.; Binversie, B.N.; Digman, M.F.; Muck, R.E.; Weimer, P.J. Single-pass, split-stream harvest of corn grain and stover. *Trans. ASABE* **2007**, *50*, 353–363.

20. Xiong, S.; Zhang, Y.; Zhuo, Y.; Lestander, T.; Geladi, P. Variations in fuel characteristics of corn (Zea mays) stovers: General spatial patterns and relationships to soil properties. *Renew. Energy* **2010**, *35*, 1185–1191.

21. Liu, J.L.; Cheng, X.; Xie, G.H.; Zhu, W.B.; Xiong, S.J. Variation in corn stover yield and fuel quality with harvest time. In Proceedings of the Asia-Pacific Power and Energy Engineering Conference, Wuhan, China, 27–31 March 2009; pp. 1–6.

22. Shinners, K.J.; Bennett, R.G.; Hoffman, D.S. Single- and two-pass corn grain and stover harvesting. *Trans. ASABE* **2012**, *55*, 341–350.

23. Hoskinson, R.L.; Karlen, D.L.; Birrell, S.J.; Radtke, C.W.; Wilhelm, W.W. Engineering, nutrient removal, and feedstock conversion evaluations of four corn stover harvest scenarios. *Biomass Bioenergy* **2007**, *31*, 126–136.

10

Influence of Coal Blending on Ash Fusibility in Reducing Atmosphere

Mingke Shen [1], Kunzan Qiu [1,*], Long Zhang [2], Zhenyu Huang [1], Zhihua Wang [1] and Jianzhong Liu [1]

[1] State Key Laboratory of Clean Energy Utilization, Zhejiang University, Hangzhou 310027, China;
    E-Mails: 21327009@zju.edu.cn (M.S.); huangzy@zju.edu.cn (Z.H.); wangzh@zju.edu.cn (Z.W.);
    jzliu@zju.edu.cn (J.L.)

[2] Laboratory for Thermal Hydraulic and Safety, China Nuclear Power Technology Research Institute,
    Shenzhen 528026, China; E-Mail: 21227081@zju.edu.com

* Author to whom correspondence should be addressed; E-Mail: qiukz@zju.edu.cn

Academic Editor: Mehrdad Massoudi

Abstract: Coal blending is an effective way to organize and control coal ash fusibility to meet different requirements of Coal-fired power plants. This study investigates three different eutectic processes and explains the mechanism of how coal blending affects ash fusibility. The blended ashes were prepared by hand-mixing two raw coal ashes at five blending ratios, G:D = 10:90 (G10D90), G:D= 20:80 (G20D80), G:D = 30:70 (G30D70), G:D = 40:60 (G40D60), and G:D = 50:50 (G50D50). The samples were heated at 900 °C, 1000 °C, 1100 °C, 1200 °C, and 1300 °C in reducing atmosphere. XRD and SEM/EDX were used to identify mineral transformations and eutectic processes. The eutectic processes were finally simulated with FactSage. Results show that the fusion temperatures of the blended ashes initially decrease and then increase with the blending ratio, a trend that is typical of eutectic melting. Eutectic phenomena are observed in D100, G10D90, and G30D70 in different degrees, which do not appear in G100 and G50D50 for the lack of eutectic reactants. The main eutectic reactants are gehlenite, magnetite, merwinite, and diopside. The FactSage simulation results show that the content discrepancy of merwinite and diopside in the ashes causes the inconsistent eutectic temperatures and eutectic degrees, in turn decrease the fusion temperature of the blended ash and then increase them with the blending ratio.

Keywords: coal blending; ash fusibility; mineral transformations; eutectic

---

1. Introduction

Coal-fired power plants have different requirements for ash fusibility. Solid-state slag-tap boilers generally require pulverized coal with a high ash fusion temperature to prevent slagging, whereas the liquid-state slag-tap boilers require a low ash fusion temperature [1]. Coal-fired power plants use the method of additive blending or coal blending to meet ash fusibility and other requirements, such as reducing slagging and ash deposition. However, additive blending is costly for plants, because additives consume coal heat and do not release heat. Coal blending has attracted significant attention because it does not increase ash content and oxygen demand, which are its obvious advantages compared with additive blending [2–4]. Therefore, it is worth further researching in understanding the ash melting mechanism of blended coal.

A number of studies have focused on coal ash fusibility [5–8]. Lolja found that oxide composition plays a more important role in fusibility than the mineral composition does, whereas Xu argued that analyzing ash fusibility only from the elements is irrational [5,9]. According to their diverse effects on ash fusibility, the minerals in coal ash are classified as refractory minerals such as quartz, metakaolin, mullite and rutile, or fluxing minerals gypsum, potash feldspar, fayalite, and almandine [10,11]. The behavior of minerals in coal ash plays a crucial role in fusibility [12–15]. Synthetic coal ash has been used to demonstrate the effect of silica-alumina ratio on ash fusibility. The ash fusion temperature increases with the silica-alumina ratio from 1.6 to 4.0 [16]. Dialing the appropriate additive can change the mineral composition in coal ash [17–19]. Ferrum-based flux has been added into coal ash, and the flow temperature falls to below 1350 °C. Ash deposition and slagging problems associated with coal burning can effectively solve with coal washing and additive blending [17]. Blends of biomass and coal also appear and become noticeable [20,21].

Research on the fusibility mechanism of coal blending has increased significantly because of the superiority of coal blending. The ASTM test, the TG and DTG methods, X-ray diffraction (XRD), FTIR, the ternary phase diagram system, and quantum chemistry calculation have been used to analyze the mineral melting behavior and mineral reaction mechanism of blended coal ash [15,22–25]. The ash fusion temperature can be higher or lower than that of individual raw coals, not change linearly with the blending ratios [23]. It is a valid approach to predict ash fusibility during the combustion of blended coal with thermodynamic modeling [2,26]. Moreover, the ash fusion temperature (AFT) of coal ash is observed lower than the predicted liquidus temperature [2]. The complicated interaction mechanism of minerals in blended coal ash at the molecular level has been investigated with the *ab initio* density functional calculation method [22]. FactSage is a computing system with a large integrated database, including large amounts of thermodynamic database on silicate and oxide. Combined with thermodynamic data, the equilibrium state of mixture can be calculated on the basis of Gibbs free energy under certain conditions, which are used in studying the thermal conversion process of minerals in coal ash [27–30]; some parameters, such as the liquidus temperature and the slag content of coal ash cannot be measured by experiment alone [30–32].

Generally, the way mineral matters behave in blended coal ash at high temperatures has yet to be fully understood. The mechanism of how coal blending affects ash fusibility requires further research. Moreover, the additive blending or coal blending can lower ash fusion temperature because a low-temperature eutectic mixture is generated [11,22,23,33]. On the contrary, the ash fusion temperature is expected to increase if a low-temperature eutectic mixture is not generated. However, limited studies explain the detailed mechanism of how the low-temperature eutectic mixture affects ash fusibility.

In this study, Zhungeer coal (high silicon aluminum) and Zhundong coal (high calcium) were chosen as the raw materials. The slag content produced during heating in coal ash was represented by the shrinkage of ash discs, which is also used to estimate the ash fusibility compared with AFT. The blended ash fusibility shows the observable characteristic of eutectic melting. The mineral transformations in coal ash with different blending ratio were analyzed with XRD and SEM/EDX. Moreover, eutectic processes were also simulated with FactSage and the eutectic temperatures were obtained. The simulation results were used to explain why the ash fusion temperatures initially decrease and then increase with the blending ratio.

2. Experimental Section

2.1. Coal and Ash Samples

Zhungeer coal (denoted as G coal) and Zhundong coal (denoted as D coal) served as the raw materials in this study. Both coal samples were primarily pulverized into fine particles with diameters less than 100 μm, and then combusted in a muffle furnace at 800 °C for 2 h. For temperatures lower than 800 °C, no chemical reactions other than decomposition occurred among the minerals of the coal ashes [34]. Therefore, the blended ashes were prepared by hand-mixing of the G coal ash with the D coal ash obtained at 800 °C at five blending ratios: G:D = 10:90 (G10D90), G:D= 20:80 (G20D80), G:D = 30:70 (G30D70), G:D = 40:60 (G40D60), and G:D = 50:50 (G50D50). For comparison, the G and D coal samples were also tested. The corresponding cases are denoted as G100 (100% G coal sample) and D100 (100% D coal sample). Table 1 shows that G coal belonging to high silicon aluminum coal, has high contents of SiO_2 and Al_2O_3, both accounting for 90.34% in the total ash mass. D coal belonging to high calcium coal has high content of CaO at 39.06%. Moreover, the contents of Na_2O and MgO in D100 are much higher than those in G100, reaching 4.08% and 9.58%, respectively.

2.2. AFT Tests

The AFTs of the samples in reducing atmosphere were calculated with an intelligence ash melting point apparatus according to GB1274–2007 standard. The ash samples were ground to 100 μm and put in a pyrometric cone, which were pushed into an ash fusion temperature auto-analyzer chamber close to a thermocouple. The heating rate was controlled at 15 to 20 °C/min under 900 °C and 5 °C/min above 900 °C in reducing atmosphere.

Table 1. Coal analysis (air-dried basis) and ash composition.

Samples	G	D
Proximate Analysis (wt %, ad)		
Moisture	6.93	13.43
Ash	24.02	5.11
Volatile matter	26.62	27.01
Fixed carbon	42.43	54.39
Ultimate Analysis (wt %, ad)		
C	54.58	64.47
H	3.31	3.11
O	9.43	12.64
N	1.13	0.69
S	0.6	0.49
Ash Composition (wt %, normalized)		
Na_2O	0.16	4.08
MgO	1.24	9.58
Al_2O_3	37.82	9.46
SiO_2	52.52	16.78
SO_3	0.51	20.38
K_2O	0.11	0.65
CaO	2.81	39.06
Fe_2O_3	3.34	3.68

2.3. Heat Treatment Experiments

A high-temperature tube furnace was used in this study and its highest heating temperature was 1700 °C. To simulate the weak reducing atmosphere, a mixture of CO_2 and CO was utilized at the blending ratio of 3:2 on a volume basis in the test. For each sample, about 0.5 g ash and the corresponding ash disc were prepared. Individual coal ashes and blended ashes were then heated at 900 °C, 1000 °C, 1100 °C, 1200 °C and 1300 °C. The heating rate was controlled at 10 °C/min under 900 °C and 5 °C/min above 900 °C. After they were heated at constant temperatures for 30 min, all samples were quenched with liquid nitrogen to avoid crystallinity changes. Under this process, the ash temperature can drop to below 300 °C in 2 s. The shrinkages of the discs were calculated with Equation (1) when the ash temperatures reached room temperature, which can represent the slag content in coal ash at high temperatures.

$$shrinkage = \frac{\left(V\left(T_0\right) - V\left(T_i\right)\right)}{V\left(T_0\right)} \times 100\% \tag{1}$$

where $V(T_0)$ is the disc volume before the heat treatment (mm³); T_i is the heat temperature °C; and $V(T_i)$ is the disc volume after the heat treatment of T_i (mm³).

2.4. XRD and SEM/EDX Analysis

The mineral composition and type at each temperature interval were determined by B/Max IIIBX-ray diffractometer (Rigaku, Tokyo, Japan) with the use of Cu target, Ni ray filter, 30 mA tube current and 30 kV tube voltage. The Rietveld method was applied for semi-quantitative analysis to determine the mineral mass fraction in ash. The heat-treated samples were also analyzed with scanning electron microscopy (FEI Quanta FEG650, FEI, Hillsboro, OR, USA) with X-ray microanalysis (SEM/EDX, FEI, Hillsboro, OR, USA). Mineral structure and the changes were measured by SEM (FEI, Hillsboro, OR, USA), and weight and atomic percentages of the elements for each analyzed mineral particle were obtained from EDX analysis (FEI, Hillsboro, OR, USA).

2.5. Thermodynamic Calculation with FactSage

2.5.1. Ternary Phase Diagram and Liquidus Temperature Calculation

Si, Al and Ca were selected as the main elements to prepare a ternary phase diagram and calculate the liquids temperature diagram with FactSage (GTT, Aachen, Germany). The positions of G100, D100, G10D90, G30D70, and G50D50 were expressed in the ternary phase diagram. The primary crystal region and ash fusion temperature were analyzed. And the calculated results were utilized to compare with AFTs obtained by experiment.

2.5.2. Slag Content and Eutectic Analysis

According to XRD analysis results, the mineral relative contents were determined with semi-quantitative analysis based on the Rietveld method, which were used to simulate the eutectic process by FactSage and calculate the slag contents. The simulations were based on five elements, Si, Al, Ca, Fe and Mg.

3. Results and Discussion

3.1. Ash Fusion Characteristics

The AFTs of the samples, including DT (deformation temperature), ST (softening temperature), HT (hemispherical temperature), and FT (flow temperature) are presented in Table 2. Both G coal and D coal belong to high ash fusion temperature coal, but the blended ashes fusion temperatures are lower than the temperature of each of the coals. The ash fusion temperatures initially decrease and then subsequently increase with the blending ratio (proportion of G100), before they reach the lowest temperatures in the vicinity of G30D70 (Figure 1). The ash fusion temperatures do not change linearly with bending ratios [23]. The fusibility of blended ashes as shown above is typically caused by eutectic melting.

Table 2. Fusion characteristic temperature of the samples °C.

Samples	DT (°C)	ST (°C)	HT (°C)	FT (°C)
G100	1490	>1500	>1500	>1500
D100	1322	1371	1390	1437
G10D90	1243	1258	1271	1291
G20D80	1211	1225	1238	1257
G30D70	1167	1173	1182	1194
G40D60	1230	1242	1263	1276
G50D50	1302	1331	1368	1391

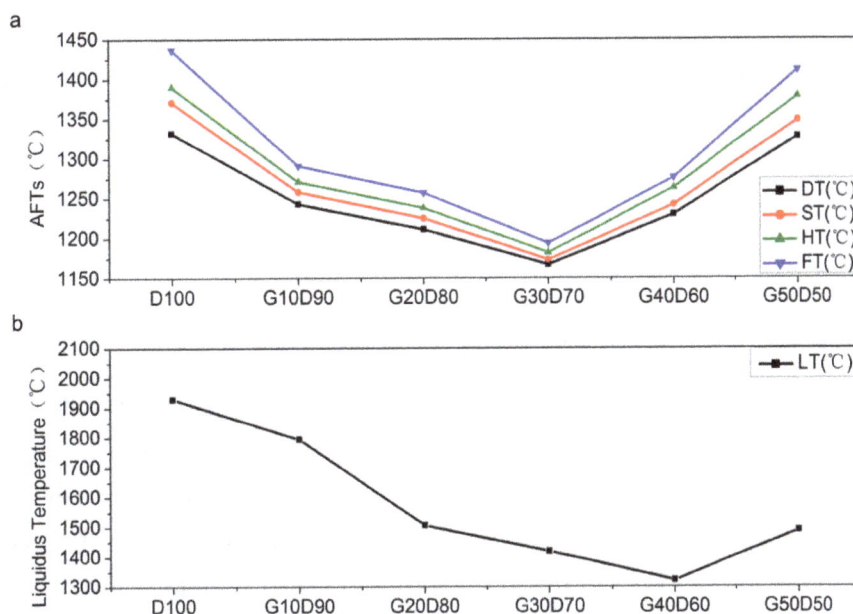

Figure 1. Comparison of ash fusion temperatures and liquid temperatures: (**a**) AFTs; (**b**) liquidus temperature.

The shrinkages of ash discs vary with temperature as showed in Figure 2. The measurements of G30D70 at 1300 °C are excluded because the disc has completely melted and cannot maintain a fixed form. This result indicates that all curves of five samples are flat at low temperatures, and the shrinkages are low, or the slag content is low. When heated to 1100 °C, the curve of G30D70 sharply ascends, and the shrinkage considerably increases, reaches 39.9%, a result indicating that large amounts of slag are generated in G30D70. At 1200 °C, the curve of G10D90 suddenly increases, particularly the amounts of slag generated in G10D90, but less dramatically than the curve of G30D70. The shrinkage of G10D90 reaches 31.2% at 1200 °C. The curve growth rate of D100 also increases at 1200 °C to 1300 °C, a result indicating that more slag generates compared with other temperatures. Moreover, the curves of G100 and G50D50 are relatively flat in the entire process at 900 °C to 1300 °C, and this means that the slag content increases slowly.

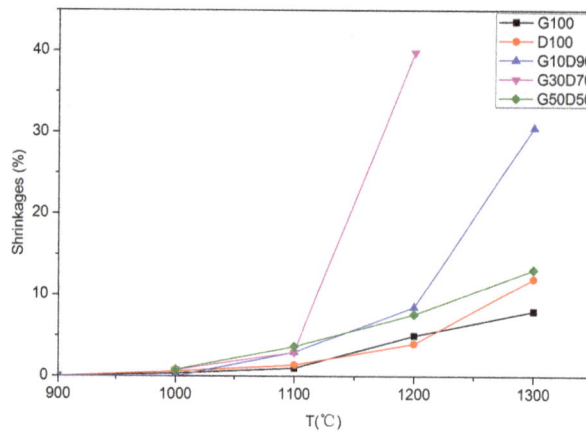

Figure 2. Shrinkages of discs made of G100, D100, G10D90, G30D70, and G50D50.

3.2. Mineral Matter Transformations of Blended Ashes

3.2.1. G100

Figure 3a presents the XRD patterns of the G100 sample for individual ash. Figure 3b shows the mineral diffraction peak intensity values of G100. Table 3 presents the mineral crystals formed in ash samples at different temperatures. The results show that no obvious diffraction peak can be observed at 900 °C, and the background diffraction is strong, a result indicating that the main components of G100 are non-crystalline substances. The melt phase cannot abound in G100 at 900 °C, so the non-crystalline substances are determined to be amorphous minerals. G100 is rich in SiO_2 and Al_2O_3 (Table 1), so the amorphous minerals are possibly clay minerals that are common in high silicon aluminum coal, such as kaolin ($xSiO_2 \cdot yAl_2O_3$). The main crystallized minerals in G100 are anhydrite and sodium silicate. The CaO and Na_2O are negligible (Table 1) so the anhydrite and sodium silicate contents are low as well. When heated at 900 °C to 1000 °C, the anhydrite diffraction peaks gradually weaken, and the sodium silicate diffraction peaks disappear. Discernible sillimanite diffraction peaks ($Al_2O_3 \cdot SiO_2$) and mullite ($3Al_2O_3 \cdot 2SiO_2$) diffraction peaks can also be observed (Figure 3a). The reactions between the minerals in G100 are as follows [17]:

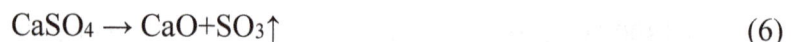

$$xSiO_2 \cdot yAl_2O_3 \rightarrow Al_2O_3 \cdot 2SiO_2 + SiO_2 \tag{2}$$

$$2(Al_2O_3 \cdot 2SiO_2) \rightarrow Al_2O_3 \text{ (amorphous)} + 3Al_2O_3 \cdot 2SiO_2 \text{ (mullite)} \tag{3}$$

$$Al_2O_3 \cdot 2SiO_2 \rightarrow Al_2O_3 \cdot SiO_2 \text{ (sillimanite)} + SiO_2 \tag{4}$$

$$2SiO_2 + Al_2O_3 \rightarrow Al_2O_3 \cdot 2SiO_2 \tag{5}$$

$$CaSO_4 \rightarrow CaO + SO_3\uparrow \tag{6}$$

Table 3. Mineral crystals formed in ash samples at different temperatures.

Mineral crystals	G100					D100					G10D90					G30D70					G50D50				
	900	1000	1100	1200	1300	900	1000	1100	1200	1300	900	1000	1100	1200	1300	900	1000	1100	1200	1300	900	1000	1100	1200	1300
Anhydrite	√					√	√	√			√	√	√			√	√				√	√	√	√	√
Sodiumsillicate	√																								
Mullite		√	√	√	√																				
Sillimanite		√	√	√																					
Cristobalite				√	√																				
Nepheline						√	√	√	√		√	√	√			√	√								
Magnetite						√	√	√	√	√	√	√	√	√		√	√	√	√		√	√	√	√	√
Gehlenite						√	√	√	√	√	√	√	√			√	√	√	√		√	√	√	√	√
Periclase						√	√	√	√		√	√	√												
Anorthite																					√	√	√	√	√
Rankinite						√																			
Merwinite							√	√	√		√	√	√	√		√	√	√	√						
Quartz						√	√				√					√	√	√	√		√	√	√	√	
Hematite																√	√	√	√		√	√	√	√	
Diopside																√	√	√	√						
Hercynite																						√	√	√	√

"√" means the corresponding mineral crystal is formed in the sample.

When heated at 1000 °C to 1100 °C, the above reactions continue. Anhydrite completely decomposes and the corresponding diffraction peaks disappear; the intensity of mullite diffraction peaks becomes stronger gradually. The SiO_2 content continues to increase because of Reaction (2). Crystal recombination is observed in the high-temperature phase, and SiO_2 is converted from an amorphous structure into a cristobalite crystal; this change leads to the appearance of cristobalite diffraction peaks at 1200 °C. Mullite diffraction peaks and sillimanite diffraction peaks also increase significantly. Because of their similar crystal structures, the sillimanite diffraction peaks are very close to the mullite diffraction peaks, phenomenon that results in peak overlapping, deformation, and smoothness. At the 1200 °C to 1300 °C temperature phase, Reaction (7) occurs, and sillimanite is converted into mullite completely.

$$Al_2O_3 \cdot SiO_2 \text{ (sillimanite)} + 2Al_2O_3 + SiO_2 \rightarrow 3Al_2O_3 \cdot 2SiO_2 \text{ (mullite)} \qquad (7)$$

Because of the disappearance of sillimanite, mullite diffraction peaks are expected to be sharp. Only mullite diffraction peaks and cristobalite diffraction peaks are observed in the diffraction pattern whose intensity is considerably increased at 1300 °C. Figure 4a shows the mullite crystal morphology found at 1300 °C in G100. For mullite and cristobalite with a high fusion temperature generated, the G100 ash fusion temperature is higher than that of the other ash samples.

3.2.2. D100

Figure 3c shows the XRD patterns of the D100 sample and Figure 3d presents the mineral diffraction peak intensity values of D100. The results show obvious nepheline ($Na_2O \cdot Al_2O_3 \cdot 2SiO_2$) diffraction peaks and periclase diffraction peaks at 900 °C, a result that is consistent with D100 being rich in Na and Mg elements (Table 1). When heated at 900 °C to 1100 °C, nepheline gradually decomposes until it disappears at 1100 °C. The most noticeable diffraction peaks are anhydrite, gehlenite and magnetite, and this observation is directly related to the high Ca element content up to 39% in D coal ash (Table 1). Amounts of anhydrite crystal were found in D100 at 900 °C, as shown in Figure 4b. In the high-temperature evolution process of abnormal calcium coal ash, the calcium minerals play a decisive role in mineral matter transformation.

The decomposition temperature of anhydrite is 900 °C to 1000 °C and the decomposition product CaO can directly react with other minerals in the coal ash [22].

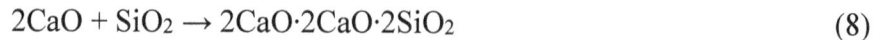

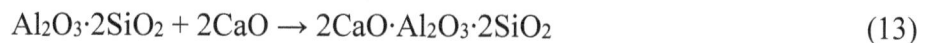

$$2CaO + SiO_2 \rightarrow 2CaO \cdot 2CaO \cdot 2SiO_2 \qquad (8)$$

$$2CaO + SiO_2 + Al_2O_3 \rightarrow 2CaO \cdot Al_2O_3 \cdot SiO_2 \text{ (gehlenite)} \qquad (9)$$

$$3CaO + SiO_2 + MgO \rightarrow 3CaO \cdot MgO \cdot 2SiO_2 \text{ (merwinite)} \qquad (10)$$

$$Al_2O_3 \cdot 2SiO_2 + 2CaO \rightarrow 2CaO \cdot Al_2O_3 \cdot SiO_2 \text{ (gehlenite)} + SiO_2 \qquad (11)$$

$$Al_2O_3 \cdot 2SiO_2 + CaO \rightarrow CaO \cdot Al_2O_3 \cdot 2SiO_2 \text{ (anorthite)} \qquad (12)$$

$$Al_2O_3 \cdot 2SiO_2 + 2CaO \rightarrow 2CaO \cdot Al_2O_3 \cdot 2SiO_2 \qquad (13)$$

Because of Reactions (9) and (11), the gehlenite diffraction peaks increase gradually. Compared with the case at 900 °C, quartz diffraction peaks considerably decrease at 1000 °C when the merwinite ($3CaO \cdot MgO \cdot 2SiO_2$) diffraction peaks appear for the first time and constantly increase because of

Reaction (10). When heated at 1200 °C to 1300 °C, the gehlenite diffraction peaks decrease, whereas the merwinite diffraction peaks and the magnetite diffraction peaks simultaneously decrease sharply and disappear at 1300 °C. The fusion temperatures of gehlenite, merwinite and magnetite are all higher than 1200 °C, so these three kinds of minerals, namely, gehlenite, merwinite, and magnetite, react and form a low-melting eutectic mixture. Finally, only gehlenite diffraction peaks and periclase diffraction peaks can be observed at 1300 °C in the XRD patterns. Both fusion temperatures of gehlenite and periclase are high, so D100 also has a high ash fusion temperature.

3.2.3. G10D90

Figure 3e presents the XRD patterns of the G10D90 sample for the blended ash. Figure 3f presents the mineral diffraction peak intensity values of G10D90. At 900 °C, the main minerals in both ashes such as nepheline, periclase, gehlenite, anhydrite, merwinite, and magnetite are similar. The introduction of G100 causes the main mineral contents in G10D90 to be lower than those in D100. It also introduces silica-alumina-abundant minerals, such as xSiO$_2$·yAl$_2$O$_3$ minerals mentioned in the previous section, into G10D90; these minerals are similar in chemical property. The sufficient SiO$_2$ is favorable to the formation of merwinite in Reaction (10). Moreover, the xSiO$_2$·yAl$_2$O$_3$ minerals react readily with the CaO generated by anhydrite and produce xSiO$_2$·yAl$_2$O$_3$·zCaO minerals, such as gehlenite, and anorthite, as shown in Reactions (9)–(12) [22].

At 900 °C to 1000 °C, the main mineral changes also include the decomposition of anhydrite and the generation of gehlenite as D100. The most significant difference between G10D90 and D100 is that gehlenite, magnetite and merwinite diffraction peaks disappear simultaneously, result indicating rapid and complete melting and conversion into a glassy state at 1200 °C to 1300 °C, so the slag content in coal ash increases noticeably (Figure 2). The fusion temperatures of gehlenite, merwinite and magnetite are all higher than 1200 °C, so gehlenite, merwinite and magnetite form a low-melting eutectic mixture, which results in gehlenite melting. The crystal morphologies of merwinite and magnetite found in G10D90 at 1200 °C are shown in Figures 4c and 4d. Although the G10D90 involves the same eutectic reactants with D100, the eutectic process is significantly different and thus accounts for the discrepancy in reactant content. The merwinite diffraction intensity of G10D90 increases considerably compared with that of D100. The content advantage results in a more violent eutectic process, which is the cause of the lower ash fusion of G10D90 than that of D100.

3.2.4. G30D70

Figure 3g presents the XRD patterns of the G30D70 sample. Figure 3h presents the mineral diffraction peak intensity values of G30D70. As more G100 is introduced, the quartz diffraction peaks considerably increase, and the main diffraction peak reaches 803 cps at 900 °C (Figure 3h); the main diffraction peak of G10D90 only reaches 438 cps (Figure 3f). Amounts of SiO$_2$ react with periclase and generate diopside, as shown in Reaction (14), so weak periclase diffraction peaks appear at 900 °C.

At 900 °C to 1100 °C, the anhydrite decomposes into CaO, which reacts to generate gehlenite and consume amounts of SiO$_2$. Because of overlapping with the nepheline and magnetite diffraction peaks, the diopside (CaO·MgO·2SiO$_2$) diffraction peaks do not appear until nepheline decomposes and decreases. Figure 3g shows that nepheline has three diffraction peaks, and two of them decrease synchronously with the temperature, whereas the rest remains the same. The increase in the diopside peaks makes up for the loss of the nepheline peaks.

$$2SiO_2 + CaO + MgO \rightarrow CaO \cdot MgO \cdot 2SiO_2 \text{ (diopside)} \qquad (14)$$

Also a result of overlapping, the nepheline diffraction peaks and the magnetite diffraction peaks become flat with the increase in diopside content. At 900 °C to 1100 °C, no main mineral disappears in G30D70, so no large amount of slag is produced in this phase, as shown in Figure 2. Only gehlenite, magnetite, and diopside diffraction peaks can be observed at 1100 °C. Amounts of diopside crystals are found in G30D70 at 1100 °C (Figure 4e). With heating at 1200 °C, no diffraction peaks in the XRD patterns are observed, so the crystallized minerals have all melted into vitresence. The high fusion temperature minerals of gehlenite, diopside and magnetite melt simultaneously, a result indicating that the minerals have formed a low-melting eutectic mixture. The eutectic mixture morphology is shown in Figure 4f. Compared with G10D90, the eutectic process obviously changes. The diopside is a substitute for merwinite as a reactant and shows an obvious advantage in terms of content. Consequently, G30D70 has the lowest ash fusion temperature among all the samples.

3.2.5. G50D50

Figure 3i presents the XRD patterns of the G50D50 sample. Figure 3j presents the mineral diffraction peak intensity values of G50D50. Compared with G30D70, more G100 is introduced to G50D50, so the proportion of the original mineral mass of D100 in G50D50, such as anhydrite, periclase, and magnetite, in the total G50D50 decreases. With the decrease in anhydrite, the CaO decomposed by anhydrite also decreases. SiO_2 and Al_2O_3 are abundant, so they inhibit Reaction (15), narrow the gehlenite and accumulate anorthite. When heated at 900 °C to 1100 °C, hematite and magnetite show no apparent change, and no new mineral is generated. When heated at 1100 °C to 1200 °C, hematite and magnetite transform into iozite (FeO), which reacts with anorthite in Reaction (16) and generates hercynite and fayalite [22]. However, fayalite melts immediately because of its low fusion temperature. Therefore, compared with the case at 1100 °C, the hematite and magnetite diffraction peaks decrease significantly at 1200 °C in the XRD patterns.

$$CaO \cdot Al_2O_3 \cdot 2SiO_2 \text{ (anorthite)} + CaO \rightarrow 2CaO \cdot Al_2O_3 \cdot 2SiO_2 \text{ (gehlenite)} \qquad (15)$$

$$CaO \cdot Al_2O_3 \cdot 2SiO_2 \text{ (gehlenite)} + FeO \rightarrow 2FeO \cdot SiO_2 + FeAl_2O_4 \qquad (16)$$

The minerals in G50D50 at 1200 °C are not evidently different from those in G50D50 at 1300 °C, and the slag content also changes very negligibly (Figure 2). In the entire temperature range, G50D50 has not shown an obvious eutectic phenomenon, probably because of the lack of periclase, and anhydrite, as well as the excess of Al_2O_3 and SiO_2. The anhydrite crystal morphology at 1300 °C in G50D50 is shown in Figure 4g. Therefore, G50D50 is given a relatively high ash fusion temperature.

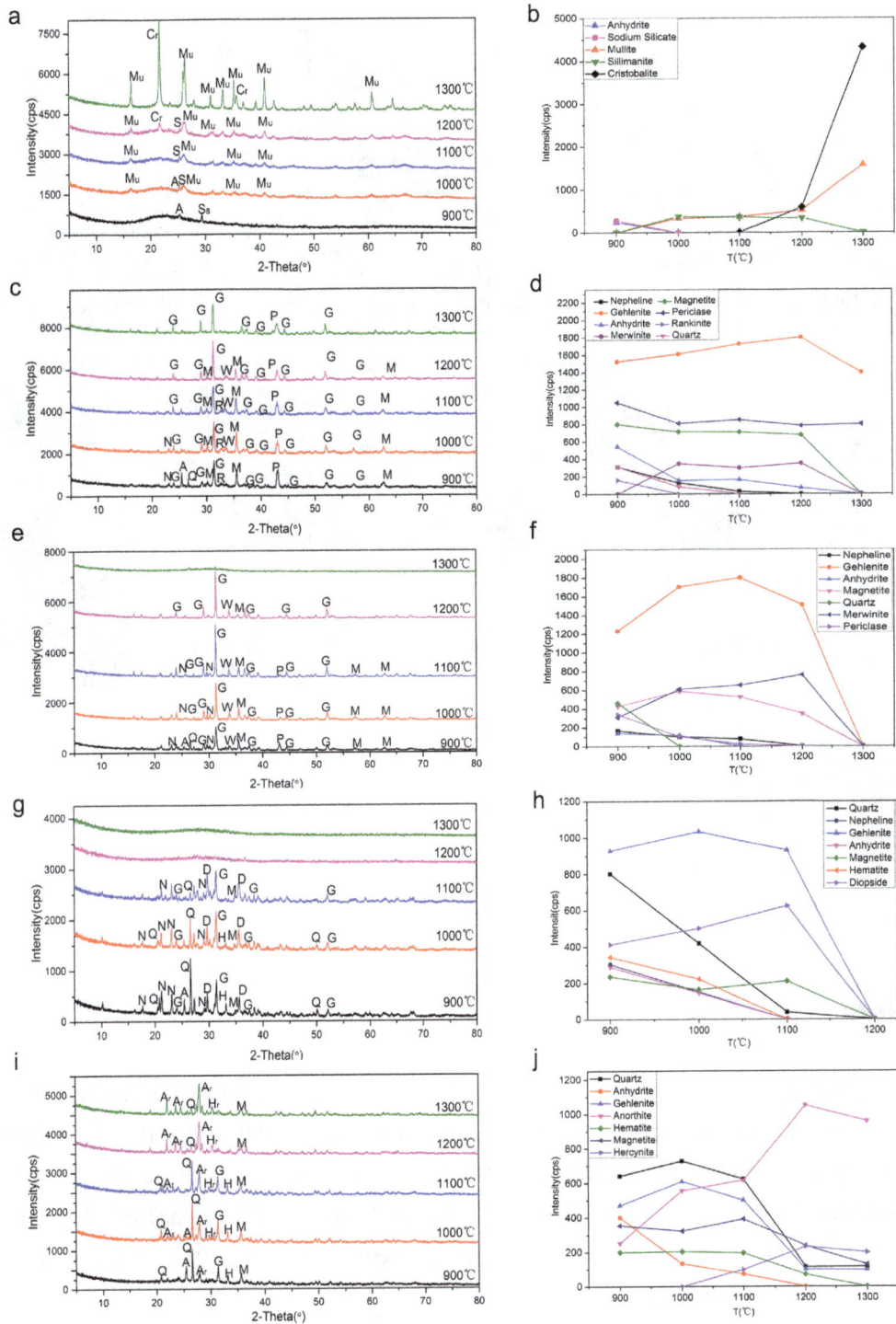

Figure 3. X-ray diffraction (XRD) patterns and intensity-temperature curves at different temperatures: (**a**) and (**b**), G100; (**c**) and (**d**), D100; (**e**) and (**f**), G10D90; (**g**) and (**h**), G30D70; and (**i**) and (**j**), G50D50. A, anhydrite; Ss, sodium silicate; Mu, mullite; S, sillimanite; Cr, cristobalite; G, gehlenite; N, nepheline; Q, quartz; P, periclase; R, rankinite; M, magnetite; W, merwinite; H, hematite; D, diopside; Hr, hercynite; Ar, anorthite and Hr, hercynite.

Figure 4. Scanning Electron Microscopy (SEM) and Energy Dispersive X-Ray (EDX) spectra of the mineral phases found in the ash samples: (**a**) crystal of mullite in G100 at 1300 °C; (**b**) crystal of anhydrite in D100 at 900 °C; (**c**) crystal of magnetite in D10G90 at 1200 °C; (**d**) morphology of merwinite in G10D90 at 1200 °C; (**e**) crystal of diopside in G30D70 at 900 °C; (**f**) morphology of the eutectic mixture in G30D70; and (**g**) crystal of anorthite in G50D50 at 1300 °C.

3.3. Thermodynamic Calculations with FactSage

3.3.1. Ternary Phase Diagram and Liquidus Analysis

Figure 5 shows the ternary phase diagram of SiO_2–Al_2O_3–CaO. Line G100–D100 stands for the chemical composition range of the investigated blended ashes. The mineral matter in the primary crystal region including point G100 is mullite, and that in the primary crystal region including point G50D50 is anorthite. It is consistent with the experimental results from Figure 3a,i. The liquidus temperature of G100 is higher than G50D50 and it is also consistent with the results of AFTs (Figure 1). However, Mineral matter in the primary crystal region including point D100 and G10D90 is α-Ca_2SiO_4 against gehlenite obtained from XRD analysis, and the liquidus temperature of D100 is higher than G100, against the AFTs results as well. Moreover, the ash sample with lowest liquidus temperature is G40D60 rather than G30D70, a sample with the lowest AFT. The reason for the inconformity between ternary phase diagram analysis and experiment results is that some important elements (such as Mg, Fe) have not been considered in the ternary phase diagram analysis. The minerals containing these elements react and form low-temperature eutectic mixtures, which significantly affect ash fusibility. Therefore, with less other elements, the ternary phase analysis result of G100 and G50D50 shows better consistency with experimental results.

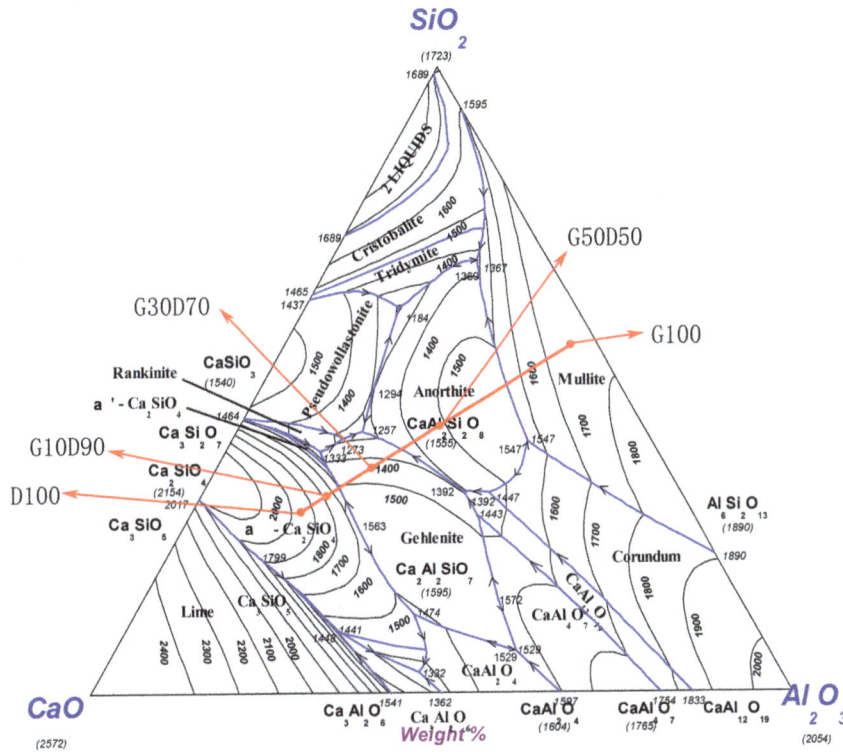

Figure 5. SiO_2–Al_2O_3–CaO ternary phase diagram.

3.3.2. Eutectic Processes Simulated with FactSage

The eutectic processes in D100, G10D90, and G30D70 were simulated with FactSage based on five elements, Si, Al, Ca, Fe and Mg, and the fusibility was analyzed according to the simulation result, which makes up for the limitation of ternary phase diagram analysis.

3.3.2.1. D100

The thermodynamic equilibrium state of gehlenite, merwinite and magnetite was calculated with FactSage at 1200 °C to 1550 °C, as shown in Figure 6a. The mineral relative mass content was obtained by semi-quantitative analysis according to the XRD analysis results (Table 4). The results indicate that when heated at 1225 °C to 1250 °C, the mass curves of merwinite, magnetite, and gehlenite descend simultaneously and sharply. The descending rates of the three mass curves at 1225 °C to 1250 °C are significantly different from those of the other temperatures, a result indicating that the descent of the mass of the three minerals has a definite correlation, namely, the emergence of the eutectic melting of the three minerals rather than individual melting or eutectic melting by twos. When the merwinite is exhausted, the eutectic process is ended. The magnetite content decreases from 100.0 g to 48.7 g and the gehlenite content decrease from 340.0 g to 316.4 g, a result indicating that just a small amount of reactants was converted into slag. The mass fraction of slag only reaches 13%, as shown in Figure 7. At 1550 °C, the coal ash has completely melted into slag. Figure 6b presents the further accurate calculation of the thermodynamic equilibrium state at 1230 °C to 1240 °C. The calculation shows that the accurate eutectic temperature of the three minerals is about 1232 °C. An accurate eutectic temperature is the characteristic of a eutectic phenomenon. The gehlenite content decreases slightly in the entire eutectic process because the merwinite content is too little to consume it. Therefore, the slag produced by the eutectic is also negligible, so the ash fusion temperature of D100 is still high.

Table 4. Relative mass content of the eutectic reactants (mass of Magnetite: 100 g).

Samples	Reactants	Mass Content (%)	Mass (g)
D100	Magnetite	21.4%	100.0
	Gehlenite	69.4%	343.0
	Merwinite	9.2%	45.3
G10D90	Magnetite	20.9%	100.0
	Gehlenite	61.2%	293.1
	Merwinite	17.9%	85.5
G30D70	Magnetite	7.9%	100.0
	Gehlenite	54.2%	470.1
	Diopside	37.9%	297.0

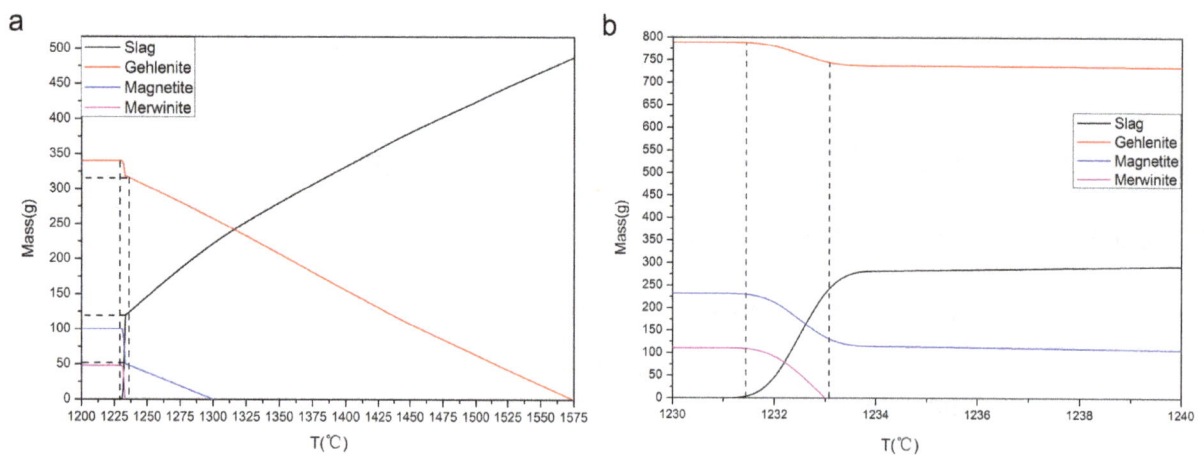

Figure 6. Mass of gehlenite, merwinite, magnetite, and slag at each thermodynamic equilibrium state (D100): (**a**) 1200 °C to 1575 °C; (**b**) 1230 °C to 1240 °C.

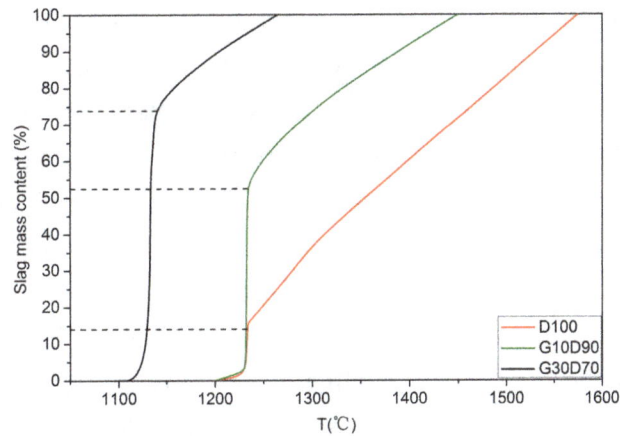

Figure 7. Comparison of the slag mass contents at each thermodynamic equilibrium state of the three blended ashes.

3.3.2.2. G10D90

The eutectic process that occurs in G10D90 is similar to the one occurs in D100, as shown above. The reactants are merwinite, magnetite and gehlenite, and the eutectic temperature is 1232 °C as well. The difference is that the mass merwinite content in G10D90 is 17.9%, which is significantly more than 9.2% in D100 (Table 4). The gehlenite content decreases from 293.1 g to 197.8 g and the magnetite decreases from 100.0 g to10.8 g (Figure 8). It means that large amounts of reactants convert into slag and the slag content reaches 55% far more than 13% in D100 during the eutectic process (Figure 7). At 1450 °C, the coal ash has completely melted into slag. Therefore, the ash fusion temperature of G10D90 is lower than that of D100.

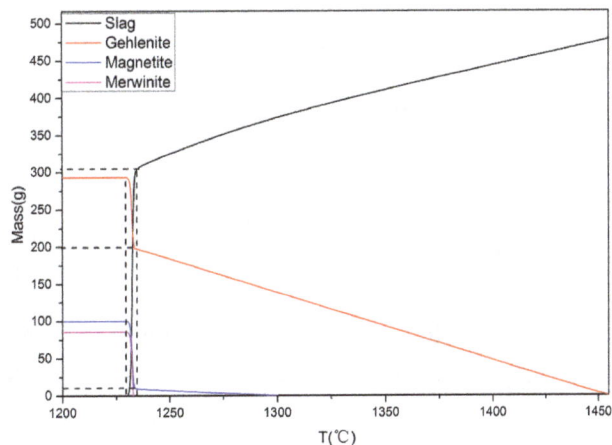

Figure 8. Masses of gehlenite, merwinite, magnetite, and slag at each thermodynamic equilibrium state (G10D90).

3.3.2.3. G30D70

Figure 9a shows the thermodynamic equilibrium state of gehlenite, diopside and magnetite at 1100 °C to 1275 °C as calculated with FactSage. As shown in Table 4, the diopside mass fraction has reached 37.9%, much higher than of merwinite in G10D90. The diopside is not directed involved in the eutectic process, but it firstly reacts with gehlenite to generate akermanite, spinel and anorthite, which consume amounts of gehlenite. Gehlenite is consumed in two ways: through reacting with diopside and though eutectic melting. When heated at 1130 °C to 1140 °C, the mass curves of gehlenite, magnetite, akermanite, spinel and anorthite decrease concurrently, a result indicating the emergence of a eutectic process. The content of gehlenite whose initial mass is 470.1 g, reaches 185.8 g after the eutectic process, a result indicating that large amounts of gehlenite were consumed and converted in slag. Figure 9b presents the further accurate calculation of the thermodynamic equilibrium state at 1124 °C to 1140 °C. The results show that the eutectic process is divided into two stages, with the disappearance of the spinel as the boundary. The first stage starts at 1132 °C, and the low-melting eutectic mixture of $Ca_2MgSi_2O_7$–$MgAl_2O_4$–$CaAl_2Si_2O_8$–$Ca_2Al_2SiO_7$–Fe_3O_4 is formed until $MgAl_2O_4$ completely melts at 1133 °C. In the second stage, the low-melting eutectic mixture of $Ca_2MgSi_2O_7$–$CaAl_2Si_2O_8$–$Ca_2Al_2SiO_7$–Fe_3O_4 is formed at 1133 °C to 1137 °C. Obviously, the mass curve of the first phase is much steeper than that of second phase. Hence, the melting performance of the former eutectic mixture is stronger than that of the latter. After the two eutectic processes, the gehlenite continue melting gradually until it completely disappears at 1275 °C. Although the eutectic phenomena also occur and reduce the gehlenite in D100 and G10D90, the eutectic temperature of G30D70 is lower about 100 °C than those of D100 and G10D90. Moreover, G30D70 has a major advantage in terms of diopside content; it can consume much more gehlenite. The slag content reaches up to 73.4% after the eutectic process (Figure 7). As a result, the fusion temperature of G30D70 is lower than those of D100 and G10D90.

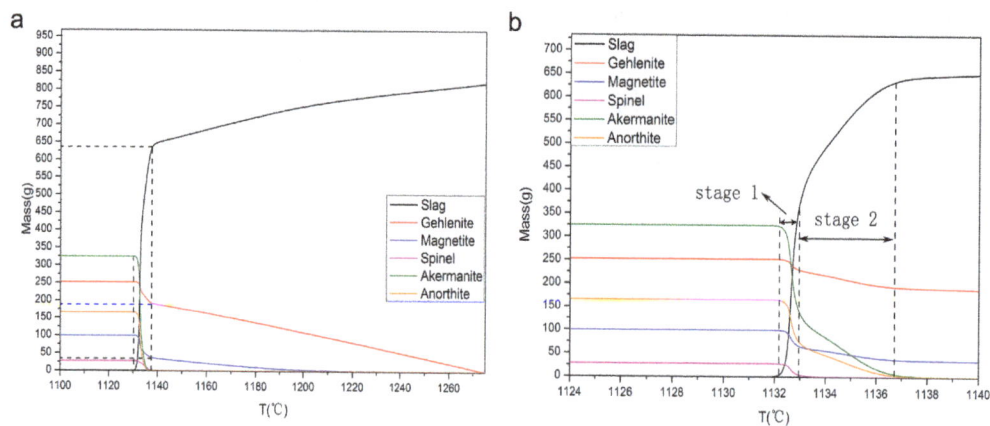

Figure 9. Masses of gehlenite, merwinite, magnetite, spinel, akermanite, anorthite, and slag at each thermodynamic equilibrium state (G30D70): **(a)** 1100 °C to 1275 °C; **(b)** 1124 °C to 1140 °C.

4. Conclusions

The ash fusion temperature of G + D blended ashes initially decreases and then increases with the blending radio, a trend that is typical of eutectic melting. The shrinkage shows a strong correlation with slag content in coal ash, and is certainly feasible to be used to represent the slag content produced during heating. Moreover, the AFTs variation tendency is not completely consistent with the results obtained by the ternary phase diagram analysis for neglecting the influence of some important elements, such as Mg and Fe.

Obvious eutectic phenomena in D100, G10D90 and G30D70 in different degrees are observed by analyzing XRD patterns and disc shrinkages, which do not appear in G100 and G50D50 because of the lack of eutectic reactants, such as merwinite and diopside. And the eutectic phenomenon is the main reason that the lower blended ash fusion temperatures than those of individual coal ash. G100 has a high ash fusion temperature because of the generation of mullite and cristobalite. The G50D50 has a relatively high ash fusion temperature because of the accumulation of anorthite. The ash fusion temperature difference between D100 and G10D90 is caused by the difference in merwinite content. Moreover, the G30D70 sample not only has a significant advantage in terms of diopside content, but the eutectic temperature of the eutectic mixture in G30D70 is also lower that of D100 and G10D90; G30D70 has the lowest ash fusion temperature among all samples.

Acknowledgments

This work was supported by the National Basic Research Program of China (2012CB214906). The authors also acknowledge the project members and may people relevant to this project.

Author Contributions

The authors Mingke Shen and Kunzan Qiu designed the research, completed the experiment and simulation, wrote and edited the paper. Long Zhang and Zhenyu Huang provided some ideas on the paper. Zhihua Wang contributed to the data collection and analysis. Jianzhong Liu checked the results and the entire manuscript.

Conflicts of Interest

The authors declare no conflict of interest.

References

1. Wang, P.; Massoudi, M. Slag Behavior in Gasifiers. Part I: Influence of Coal Properties and Gasification Conditions. *Energies* **2013**, *6*, 784–806.
2. Bryant, G.W.; Browning, G.J.; Emanuel, H.; Gupta, S.K.; Gupta, R.P.; Lucas, J.A.; Wall, T.F. The fusibility of blended coal ash. *Energy Fuels* **2000**, *14*, 316–325.
3. Sarkar, P.; Mukherjee, A.; Sahu, S.G.; Choudhury, A.; Adak, A.K.; Kumar, M.; Choudhury, N.; Biswas, S. Evaluation of combustion characteristics in thermogravimetric analyzer and drop tube furnace for Indian coal blends. *Appl. Therm. Eng.* **2013**, *60*, 145–151.
4. Zhang, X.M.; Liu, Y.H.; Wang, C.A.; Che, D.F. Experimental study on interaction and kinetic characteristics during combustion of blended coals. *J. Therm. Anal. Calorim.* **2012**, *107*, 935–942.
5. Lolja, S.A.; Haxhi, H.; Dhimitri, R.; Drushku, S.; Malja, A. Correlation between ash fusion temperatures and chemical composition in Albanian coal ashes. *Fuel* **2002**, *81*, 2257–2261.
6. Van Dyk, J.C. Understanding the influence of acidic components (Si, Al, and Ti) on ash flow temperature of South African coal sources. *Miner. Eng.* **2006**, *19*, 280–286.
7. Liu, B.; He, Q.H.; Jiang, Z.H.; Xu, R.F.; Hu, B.X. Relationship between coal ash composition and ash fusion temperatures. *Fuel* **2013**, *105*, 293–300.

8. Song, W.J.; Tang, L.H.; Zhu, X.D.; Wu, Y.Q.; Rong, Y.Q.; Zhu, Z.B.; Koyama, S. Fusibility and flow properties of coal ash and slag. *Fuel* **2009**, *88*, 297–304.

9. Xu, J.; Liu, X.; Zhao, F.; Wang, F.C.; Guo, Q.H.; Yu, G.S. Study on Fusibility and Flow Behavior of High-Calcium Coal Ash. *J. Chem. Eng. Jpn.* **2014**, *47*, 711–716.

10. Gupta, S.; Dubikova, M.; French, D.; Sahajwalla, V. Characterization of the origin and distribution of the minerals and phases in metallurgical cokes. *Energy Fuels* **2007**, *21*, 303–313.

11. Vassileva, C.G.; Vassilev, S.V. Behaviour of inorganic matter during heating of Bulgarian coals: 2. Subbituminous and bituminous coals. *Fuel Process Technol* **2006**, *87*, 1095–1116.

12. Rodrigues, S.; Marques, M.; Ward, C.R.; Suarez-Ruiz, I.; Flores, D. Mineral transformations during high temperature treatment of anthracite. *Int. J. Coal Geol.* **2012**, *94*, 191–200.

13. Chakravarty, S.; Mohanty, A.; Banerjee, A.; Tripathy, R.; Mandal, G.K.; Basariya, M.R.; Sharma, M. Composition, mineral matter characteristics and ash fusion behavior of some Indian coals. *Fuel* **2015**, *150*, 96–101.

14. Saikia, B.K.; Wang, P.P.; Saikia, A.; Song, H.J.; Liu, J.J.; Wei, J.P.; Gupta, U.N. Mineralogical and Elemental Analysis of Some High-Sulfur Indian Paleogene Coals: A Statistical Approach. *Energy Fuels* **2015**, *29*, 1407–1420.

15. Li, F.H.; Huang, J.J.; Fang, Y.T.; Liu, Q.R. Fusibility Characteristics of Residual Ash from Lignite Fluidized-Bed Gasification to Understand Its Formation. *Energy Fuels* **2012**, *26*, 5020–5027.

16. Song, W.J.; Tang, L.H.; Zhu, X.D.; Wu, Y.Q.; Zhu, Z.B.; Koyama, S. Prediction of Chinese Coal Ash Fusion Temperatures in Ar and H_2 Atmospheres. *Energy Fuels* **2009**, *23*, 1990–1997.

17. Huang, Z.Y.; Li, Y.; Lu, D.; Zhou, Z.J.; Wang, Z.H.; Zhou, J.H.; Cen, K.F. Improvement of the Coal Ash Slagging Tendency by Coal Washing and Additive Blending with Mullite Generation. *Energy Fuels* **2013**, *27*, 2049–2056.

18. Folgueras, M.B.; Alonso, M.; Folgueras, J.R. Modification of lignite ash fusion temperatures by the addition of different types of sewage sludge. *Fuel Process. Technol.* **2015**, *131*, 348–355.

19. Nel, M.V.; Strydorn, C.A.; Schobert, H.H.; Beukes, J.P.; Bunt, J.R. Reducing atmosphere ash fusion temperatures of a mixture of coal-associated minerals—The effect of inorganic additives and ashing temperature. *Fuel Process. Technol.* **2014**, *124*, 78–86.

20. Vamvuka, D.; Kakaras, E. Ash properties and environmental impact of various biomass and coal fuels and their blends. *Fuel Process. Technol.* **2011**, *92*, 570–581.

21. Kocabas-Atakli, Z.O.; Okyay-Oner, F.; Yurum, Y. Combustion characteristics of Turkish hazelnut shell biomass, lignite coal, and their respective blends via thermogravimetric analysis. *J. Therm. Anal. Calorim.* **2015**, *119*, 1723–1729.

22. Wu, X.J.; Zhang, Z.X.; Chen, Y.S.; Zhou, T.; Fan, J.J.; Piao, G.L.; Kobayashi, N.; Mori, S.; Itaya, Y. Main mineral melting behavior and mineral reaction mechanism at molecular level of blended coal ash under gasification condition. *Fuel Process. Technol.* **2010**, *91*, 1591–1600.

23. Qiu, J.R.; Li, F.; Zheng, Y.; Zheng, C.G.; Zhou, H.C. The influences of mineral behaviour on blended coal ash fusion characteristics. *Fuel* **1999**, *78*, 963–969.

24. Bai, J.; Li, W.; Li, B.Q. Characterization of low-temperature coal ash behaviors at high temperatures under reducing atmosphere. *Fuel* **2008**, *87*, 583–591.

25. Wang, C.A.; Liu, Y.H.; Zhang, X.M.; Che, D.F. A Study on Coal Properties and Combustion Characteristics of Blended Coals in Northwestern China. *Energy Fuels* **2011**, *25*, 3634–3645.

26. Djordjevic, D.; Stojkovic, D.; Djordjevic, N.; Smelcerovic, M. Thermodynamics of Reactive Dye Adsorption from Aqueous Solution on the Ashes from City Heating Station. *Ecol. Chem. Eng. S.* **2011**, *18*, 527–536.

27. Song, W.J.; Sun, Y.M.; Wu, Y.Q.; Zhu, Z.B.; Koyama, S. Measurement and Simulation of Flow Properties of Coal Ash Slag in Coal Gasification. *AICHE J.* **2011**, *57*, 801–818.

28. Kong, L.X.; Bai, J.; Bai, Z.Q.; Guo, Z.X.; Li, W. Effects of $CaCO_3$ on slag flow properties at high temperatures. *Fuel* **2013**, *109*, 76–85.

29. Zhang, G.J.; Reinmoller, M.; Klinger, M.; Meyer, B. Ash melting behavior and slag infiltration into alumina refractory simulating co-gasification of coal and biomass. *Fuel* **2015**, *139*, 457–465.

30. Ilyushechkin, A.Y.; Hla, S.S. Viscosity of High-Iron Slags from Australian Coals. *Energy Fuels* **2013**, *27*, 3736–3742.

31. Van Dyk, J.C.; Keyser, M.J. Influence of discard mineral matter on slag-liquid formation and ash melting properties of coal—A FACTSAGE (TM) simulation study. *Fuel* **2014**, *116*, 834–840.

32. Yuan, H.P.; Liang, Q.F.; Gong, X. Crystallization of Coal Ash Slags at High Temperatures and Effects on the Viscosity. *Energy Fuels* **2012**, *26*, 3717–3722.

33. Zhu, Y.J.; Piotrowska, P.; van Eyk, P.J.; Bostrom, D.; Kwong, C.W.; Wang, D.B.; Cole, A.J.; de Nys, R.; Gentili, F.G.; Ashman, P.J. Cogasification of Australian Brown Coal with Algae in a Fluidized Bed Reactor. *Energy Fuels* **2015**, *29*, 1686–1700.

34. Srinivasachar, S.; Helble, J.J.; Boni, A.A. Mineral Behavior during Coal Combustion 1. Pyrite Transformations. *Prog. Energy Combust.* **1990**, *16*, 281–292.

Permissions

All chapters in this book were first published in Energies, by MDPI; hereby published with permission under the Creative Commons Attribution License or equivalent. Every chapter published in this book has been scrutinized by our experts. Their significance has been extensively debated. The topics covered herein carry significant findings which will fuel the growth of the discipline. They may even be implemented as practical applications or may be referred to as a beginning point for another development.

The contributors of this book come from diverse backgrounds, making this book a truly international effort. This book will bring forth new frontiers with its revolutionizing research information and detailed analysis of the nascent developments around the world.

We would like to thank all the contributing authors for lending their expertise to make the book truly unique. They have played a crucial role in the development of this book. Without their invaluable contributions this book wouldn't have been possible. They have made vital efforts to compile up to date information on the varied aspects of this subject to make this book a valuable addition to the collection of many professionals and students.

This book was conceptualized with the vision of imparting up-to-date information and advanced data in this field. To ensure the same, a matchless editorial board was set up. Every individual on the board went through rigorous rounds of assessment to prove their worth. After which they invested a large part of their time researching and compiling the most relevant data for our readers.

The editorial board has been involved in producing this book since its inception. They have spent rigorous hours researching and exploring the diverse topics which have resulted in the successful publishing of this book. They have passed on their knowledge of decades through this book. To expedite this challenging task, the publisher supported the team at every step. A small team of assistant editors was also appointed to further simplify the editing procedure and attain best results for the readers.

Apart from the editorial board, the designing team has also invested a significant amount of their time in understanding the subject and creating the most relevant covers. They scrutinized every image to scout for the most suitable representation of the subject and create an appropriate cover for the book.

The publishing team has been an ardent support to the editorial, designing and production team. Their endless efforts to recruit the best for this project, has resulted in the accomplishment of this book. They are a veteran in the field of academics and their pool of knowledge is as vast as their experience in printing. Their expertise and guidance has proved useful at every step. Their uncompromising quality standards have made this book an exceptional effort. Their encouragement from time to time has been an inspiration for everyone.

The publisher and the editorial board hope that this book will prove to be a valuable piece of knowledge for researchers, students, practitioners and scholars across the globe.

List of Contributors

Piotr Kolasiński
Department of Thermodynamics, Theory of Machines and Thermal Systems, Faculty of Mechanical and Power Engineering, Wrocław University of Technology, Wybrzeże Wyspiańskiego 27, 50-370 Wrocław, Poland

Yingning Qiu
School of Energy and Power Engineering, Nanjing University of Science and Technology, Nanjing 210094, China

Wenxiu Zhang
School of Energy and Power Engineering, Nanjing University of Science and Technology, Nanjing 210094, China

Mengnan Cao
School of Energy and Power Engineering, Nanjing University of Science and Technology, Nanjing 210094, China

Yanhui Feng
School of Energy and Power Engineering, Nanjing University of Science and Technology, Nanjing 210094, China

David Infield
Department of Electronic and Electrical Engineering, Strathclyde University, Glasgow G1 1XQ, UK

Xiaowei Feng
Key Laboratory of Deep Coal Resource Mining, Ministry of Education, China University of Mining and Technology, Xuzhou 221116, China
State Key Laboratory of Coal Resources and Safe Mining, School of Mines, China University of Mining and Technology, Xuzhou 221116, China

Nong Zhang
Key Laboratory of Deep Coal Resource Mining, Ministry of Education, China University of Mining and Technology, Xuzhou 221116, China
State Key Laboratory of Coal Resources and Safe Mining, School of Mines, China University of Mining and Technology, Xuzhou 221116, China
Hunan Key Laboratory of Safe Mining Techniques of Coal Mines

Lianyuan Gong
Key Laboratory of Coal Processing and Efficient Utilization of Ministry of Education, School of Chemical Engineering and Technology, China University of Mining and Technology, Xuzhou 221116, China

Fei Xue
Key Laboratory of Deep Coal Resource Mining, Ministry of Education, China University of Mining and Technology, Xuzhou 221116, China
State Key Laboratory of Coal Resources and Safe Mining, School of Mines, China University of Mining and Technology, Xuzhou 221116, China

Xigui Zheng
Key Laboratory of Deep Coal Resource Mining, Ministry of Education, China University of Mining and Technology, Xuzhou 221116, China
State Key Laboratory of Coal Resources and Safe Mining, School of Mines, China University of Mining and Technology, Xuzhou 221116, China

Insoo Ye
School of Mechanical Engineering, Sungkyunkwan University, Suwon 440-746, Korea

Junho Oh
School of Mechanical Engineering, Sungkyunkwan University, Suwon 440-746, Korea

Changkook Ryu
School of Mechanical Engineering, Sungkyunkwan University, Suwon 440-746, Korea

Patrick Linke
Department of Chemical Engineering, Texas A&M University at Qatar, P.O. Box 23874, Education City, 77874 Doha, Qatar

Athanasios I. Papadopoulos
Chemical Process and Energy Resources Institute, Centre for Research and Technology-Hellas, Thermi, 57001 Thessaloniki, Greece

Panos Seferlis
Department of Mechanical Engineering, Aristotle University of Thessaloniki, P.O. Box 484, 54124 Thessaloniki, Greece

Akihiro Hachikubo
Environmental and Energy Resources Research Center, Kitami Institute of Technology, 165 Koen-cho, Kitami 090-8507, Japan

Katsunori Yanagawa
Faculty of Social and Cultural Studies, Kyushu University, 744 Motooka, Nishi-ku, Fukuoka 819-0395, Japan

Hitoshi Tomaru
Department of Earth Sciences, Graduate School of Science, Chiba University, 1-33 Yayoi-cho, Inageku, Chiba 263-8522, Japan

Hailong Lu
Department of Energy and Resource Engineering, College of Engineering, Peking University, Beijing 100871, China

Ryo Matsumoto
Gas Hydrate Laboratory, Organization for the Strategic Coordination of Research and Intellectual Properties, Meiji University, 1-1 Kanda-Surugadai, Chiyoda-ku, Tokyo 101-8301, Japan

Cheng Lin
Collaborative Innovation Center of Electric Vehicles in Beijing, Beijing Institute of Technology, Beijing 100081, China

Zhifeng Xu
Collaborative Innovation Center of Electric Vehicles in Beijing, Beijing Institute of Technology, Beijing 100081, China

Li-Wei Shi
Aero-Power Science Technology Center, Nanjing University of Aeronautics and Astronautics, Nanjing 210016, China

Bo Zhou
Aero-Power Science Technology Center, Nanjing University of Aeronautics and Astronautics, Nanjing 210016, China

Pierre-Luc Lizotte
Département des sols et de génie agroalimentaire, Université Laval, 2425 rue de l'Agriculture, Québec City, QC G1V 0A6, Canada

Philippe Savoie
Département des sols et de génie agroalimentaire, Université Laval, 2425 rue de l'Agriculture, Québec City, QC G1V 0A6, Canada
Agriculture and Agri-Food Canada, 2560 Hochelaga Blvd., Québec City, QC G1V 2J3, Canada

Alain De Champlain
Département de génie mécanique, Université Laval, 1065 avenue de la Médecine, Québec City, QC G1V 0A6, Canada

Mingke Shen
State Key Laboratory of Clean Energy Utilization, Zhejiang University, Hangzhou 310027, China

Kunzan Qiu
State Key Laboratory of Clean Energy Utilization, Zhejiang University, Hangzhou 310027, China

Long Zhang
Laboratory for Thermal Hydraulic and Safety, China Nuclear Power Technology Research Institute, Shenzhen 528026, China

Zhenyu Huang
State Key Laboratory of Clean Energy Utilization, Zhejiang University, Hangzhou 310027, China

Zhihua Wang
State Key Laboratory of Clean Energy Utilization, Zhejiang University, Hangzhou 310027, China

Jianzhong Liu
State Key Laboratory of Clean Energy Utilization, Zhejiang University, Hangzhou 310027, China